Molecular Biology in Ayurveda Series

Elaborated laghu guṇa

(Volume 2)

By

Dr Kulkarni Dhananjay Vyankatesh
MD, PhD (Chikitsa)
AVP, PhD (Ayurved)
MA (Sanskrit)

2019

Dedicated to
My beloved wife Dr Mrudul,
And our lovely children
Pallavi and Pranav
And
Grand-son Rutvija

Table of Contents : -

गुणज्ञानप्रस्थान

Methodology of guṇajñāna (गुणज्ञान): -

While labeling or accrediting any of the *rukṣa snigdhādi* (रुक्ष स्निग्धादि) attributes on an entity, it is very important to understand some of the concepts behind it. First important one is that the *guṇa* (गुण) or attribute needs a *āśraya* (आश्रय) or shelter or support for its manifestation and this is provided by the *dravya* (द्रव्य). In addition, *caraka* (चरक) has also mentioned that without *dravya* (द्रव्य), *guṇāḥ* (गुणाः) are unable to manifest themselves.

गुणा गुणाश्रया नोक्ताः तस्माद्रसगुणान् भिषक् । चरक सूत्र २६-३६

śrī cakrapāṇidatta (श्री चक्रपाणिदत्त) also in his commentary says that potentials do not stay in themselves. These attributes need an entity to harbor. In this way, *rukṣasūkṣmādi* (रुक्षसूक्ष्मादि) qualities are available in the *dravya* (द्रव्य). This perception is very precisely described in many places in the ancient texts. For example, while describing the qualities of *rasa* (रस), it is mentioned that these *rukṣa-snigdhādi* (रुक्ष-स्निग्धादि) attributes do not belong to *rasa* (रस), but they are of the supporting dravya because *guṇāḥ* (गुणाः) or attributes are unable to reside on other *guṇāḥ* (गुणाः).

गुर्वादयो गुणा द्रव्ये पृथिव्यादौ रसाश्रये।

रसेषु व्यपदिश्यन्ते साहचर्योपचारतः॥ वाग्भट्ट सूत्र ९-४

Some of the attributes are not visible, the meaning is that they are in existence, but are beyond the capability of the *jñāneṁdriya* (ज्ञानेंद्रिय) to acknowledge them or we can say that, they cannot be assessed by *pratyakṣa pramāṇa* (प्रत्यक्ष प्रमाण).

We know that,

a) *pratyakṣa pramāṇa* (प्रत्यक्ष प्रमाण),
b) *anumāna pramāṇa* (अनुमान प्रमाण),
c) *yukti pramāṇa* (युक्ति प्रमाण), and
d) *āptopadeśa pramāṇa* (आप्तोपदेश प्रमाण)

are the widely used *pramāṇāḥ* (प्रमाणाः), which are employed for the perception of knowledge. As discussed earlier, some attributes are beyond the human capacity of perception, so we are unable to acknowledge it by our sense organs. Therefore, the remaining three *pramāṇāḥ* (प्रमाणाः) are used for the understanding of the entities. It is well said that "*kāryānumeyaḥ guṇāḥ guṇānumeyaḥ dravyāḥ* (**कार्यानुमेयः गुणाः गुणानुमेयः द्रव्याः**)", means after perceiving the performance, action or function, we are capable to determine the characteristics or *guṇa* (गुण) responsible for that action, and bearing in mind these attributes or *guṇa* (गुण), we can make an assumption or guesswork of that *dravya* (द्रव्य), which is giving *āśraya* (आश्रय) or patronage to these attributes.

So when we want to understand the functionality of an attribute, we must look in the details of that *dravya* (द्रव्य), which is giving *āśraya* (आश्रय) or support for its manifestation. The easiest way to cognize the *dravya* (द्रव्य), is to look in its *pāṁcabhautika* (पांचभौतिक) constitution. This can be done in two ways.

1. One is to consume the *dravya* (द्रव्य) and then observe its effects on the physiological activities of the body by the ways of perceiving the *rasa* (रस), *vīrya* (वीर्य), *vipāka* (विपाक), and *karma* (कर्म). As discussed earlier, "*kāryānumeyaḥ guṇāḥ guṇānumeyaḥ dravyāḥ* (कार्यानुमेयः गुणाः गुणानुमेयः द्रव्याः)", here it is imaginable to determine the *pāṁcabhautika*

(पांचभौतिक) constitution of the dravya with the help of the activities observed.

2. Secondly, we can directly study the *dravya* (द्रव्य) for its *pāṁcabhautika* (पांचभौतिक) constitution. *Kavirāja Sri Ūpendranātha dāsa* (कविराज श्री उपेन्द्रनाथ दास) in his "*tridoṣa vimarśa* (त्रिदोष विमर्श)" has cited that the attributes like *rukṣa* (रुक्ष), *laghu* (लघु), *śīta* (शीत), *Khara* (खर), *sūkṣma* (सूक्ष्म), *cala* (चल), *viśada* (विशद), *sthira* (स्थिर), *guru* (गुरु), *snigdha* (स्निग्ध) etc., which inherent in the *doṣa* (दोष), *dhātu* (धातु), and *mala* (मल), are nourished by the same *rukṣa-sukṣmādi* (रुक्ष-सुक्ष्मादि) attributes, which are also existent in the food and drugs.

शास्त्रेषु वातादीनां ये गुणा रुक्षतोष्णतास्निग्धतादयो वर्णितास्ते भोज्यपेयादिषु अपि दृश्यन्ते। कविराज उपेन्द्रनाथ दास त्रिदोष विमर्श -१९८२ सप्तम अध्याय पृष्ठ ७१

As we are aware of the fact that, the process of growth in the body is always dependent on the concept of *vṛdhdiḥ samānaiḥ sarveṣām* (वृध्दिः समानैः सर्वेषाम्). The meaning of this verse is that application of such types of attributes from food or drugs, which are the same or equal in the attributes of the body material, always increases the bodily entities.

वृद्धिः समानैः सर्वेषम् विपरितम् विपर्ययैः ।अष्टांग हृदय सूत्रस्थान

For example, when we believe that *māṁsaṁ bṛhaṇīyānāṁ śreṣṭham* (**मांसं बृंहणीयानां श्रेष्ठम् । चरक सूत्रस्थान २५-४०**), we have admitted that, the ingested meat material is going to be utilized for the quantitative and qualitative enrichment of muscle tissues in the body.

बृंहणम् पृथिव्यम्बुगुणभुयिष्टम्। सुश्रुत सूत्रस्थान ४१-६

As per the ayurvedic concepts, the *pāaṁcabhautika* (पांचभौतिक) constitution of meat, mainly its *pārthiva* (पार्थिव) and *jaliya* (जलिय) parts, nourishes the *māṁsa dhātu* (मांस धातु) in the body.

नातिशीतं गुरु स्निग्धं मांसमाजमदोषलम्।

शरीरधातुसामान्यादनभिष्यन्दि बृंहणम्॥ अष्टांग हृदय सूत्र 6-62

We also know that the *māṁsa* (मांस) consists mainly of high proteins and after digestion these entities are transformed into large biological molecules comprising of one or more chains of amino acids. It is well established that, all muscle tissues are very high in protein, which are actually one or more chains of amino acids. So when we decide that the stability of the body is due to the *māṁsa dhātu* (मांस धातु), we are actually saying that the muscle tissues are in possession of *sthira* (स्थिर), *guru* (गुरु), *snigdha* (स्निग्ध) *guṇāḥ* (गुणाः). If we question ourselves that, from where these *sthirādī guṇa* (स्थिरादी गुण) manifested in the *māṁsa dhātu* (मांस धातु) in the body, we will observe that these *sthirādī guṇa* (स्थिरादी गुण) are also present in the *māṁsa dhātu* (मांस धातु) of the mammals.

सृष्टीविकास क्रम:

Understanding sṛṣṭīvikāsa krama (सृष्टीविकास क्रम) and manifestation of *guṇāḥ* (*गुणाः*): -

So when we consider about the understanding of the *guṇāḥ* (गुणाः), it is obvious to study these *bāhya pāṁcabhautika dravyāḥ* (बाह्य पांचभौतिक द्रव्याः) which are being consumed as food materials.

To continue the "**मांसं बृंहणीयानां श्रेष्ठम्** ।" example mentioned above, we should go further in the details of the *bāhya pāṁcabhautika dravyas* (बाह्य पांचभौतिक द्रव्य), which nourishes the *māṁsa dhātu* (मांस धातु) of the body. We have stated earlier that *māṁsa dhātu* (मांस धातु) made available from the animals are rich in proteins and proteins are made of the amino-acid chains. Means it is quite clear that, some of the amino acids should be in possession of the *sthira* (स्थिर), *guru* (गुरु), *snigdha* (स्निग्ध) *guṇāḥ* (गुणाः). As all the manifested dravyāḥ (द्रव्याः) are *pāṁcabhautika* (पांचभौतिक) in their constitution, we can conclude that such amino-acids, which are responsible for the nourishment of the *māṁsa dhātu* (मांस धातु) in the body, should be manifested by those *dravyāḥ* (द्रव्याः), which are in possession of *pṛthvi mahābhuta* (पृथ्वि महाभुत) and *jala mahābhuta* (जल महाभुत) in their *pāṁcabhautika* (पांचभौतिक) constitution.

One more thing is that these amino-acids are also formed by a specific arrangement of atoms. Actually carbon (C), hydrogen (H), oxygen (O), nitrogen (N) and sulfur (S) are the five key elements of an amino acid. But it also should be noted that some other elements are similarly found in the side chains of certain amino acids. One should never forget that these atoms also are *pāṁcabhautika* (पांचभौतिक) in their constitution.

Such conclusion leads us towards the physical, biophysical, phyto-chemical, biological, macro and electron-microscopic molecular studies of the amino-acid structure and their basic atoms, for the purpose of understanding different attribute residing in them.

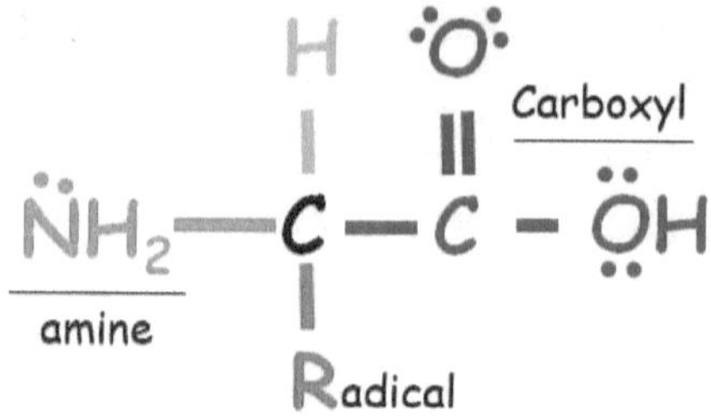

Amino Acid basic structure

Fundamentally the knowledge of more than 500 naturally occurring amino acids, is available currently. However, out of these, only 20 amino-acids express their presence in the genetic code. Out of these 20 amino-acids, nine amino-acids are labeled as essential or indispensable amino-acids, because the human physiological system is unable to synthesize them in the body at a rate corresponding with their requirement. Consequently, these should be supplied through the way of food. Essential amino-acids are phenylalanine, valine, threonine, tryptophan, methionine, leucine, isoleucine, lysine, and histidine.

Amino-acids are actually classified in different ways, but I would like to prefer a particular classification for taking into consideration, as it is more prone to provide effective information, which can be used to reveal their *pāṁcabhautika* (पांचभौतिक) constitution. This classification is based on their own structure and the structure of their side chains, and provides two basic subcategories.

1. Non-Polar amino acids.
2. Polar amino acids.

Non-Polar amino acids are in possession of equal number of amino and carboxyl groups, hence they are nutral. This group is also recognized as hydrophobic amino acids. Actually hydro means water and phobic means fearing. Dictionary meaning of the word hydrophobic is 'tending to repel or fail to mix with water'. Why water ?, because it is a highly polar substance. One more thing is that hydrophobic amino acids possess only these variable R-groups within themselves, which are composed of mostly hydrocarbons, and there is no charge on these R-groups. This phenomenon compels hydrophobic amino acids to be with little or no polarity in their side chains. The absence of polarity means, these amino acids have no way to act together with extremely polar water molecules, resulting in production of water-fearing attributes.

So this classification allows us to make 2 groups of amino acids: -

1. one which has a tendency to interact with water or mix with water and
2. another one is that which has a tendency to repel away from the water.

Therefore, when we say that hydrophobic amino acids tend to run away from the water molecules, the meaning is that there are such attributes in these hydrophobic amino acids, which are opposite in nature of the water. Non-polarity is like only the tip of the iceberg.

Apart from non-polarity, there are many attributes like : -

dryness, aridness, adust or sunburn appearance, harsh, rough, parched, dehydrated, moistureless, unpleasantly rough, hoarse or raspy, strident, not smooth, uneven or irregular surface, as dry as a bone, withered or dry and shrivelled, shrunken or wrinkled from age, contract due to loss of moisture, seared or fried or burn the surface with a sudden intense heat, desiccated or having had all moisture removed, dried out, etc.,

in which we can observe different shades of expression, where the entity is moved away from the water.

We can divide all the substances in the world into two groups,

a) first such entities which interact strongly with water, and
b) second those entities which tend to repel from water or fail to mix with water.

This concept has a very long history. Since the Vedic time, which is approximately a minimum of 5000 years B.C., the ancient evolution theory is celebrated. While elaborating the concept of brahmavidyā (ब्रह्मविद्या), an amusing description of sṛṣṭīkrama (सृष्टीक्रम) is available in taittirīyopaniṣad (तैत्तिरीयोपनिषद्), which reveals the facts that, first ākāśa (आकाश) was evoluted from the ātma svarupa brahma (आत्म स्वरुप ब्रह्म). Subsequently, vāyu (वायु) was created from this ākāśa (आकाश), and in this theory of evolution, it is said that, after the formation of vāyu mahābhuta (वायु महाभुत), it gives birth to the agnī mahābhuta (अग्नी महाभुत), which is mentioned in the original verse as “vāyoragniḥ (वायोरग्निः)”. After the formation of agnī mahābhuta (अग्नी महाभुत), it gives birth to the jala mahābhuta (जल महाभुत), which becomes responsible for the birth of pṛthvi mahābhuta (पृथ्वि महाभुत) afterward.

(Taittiriyopanishad Shankarbhashya sahit / Brahmanandvalli / Pratham anuwak / 1 / page 84; Gita press Gorakhpur Savatsa 2064.)

As per the sṛṣṭīkrama (सृष्टीक्रम) mentioned above, until the materialization of agni, there was no existence of jala mahābhuta. Meaning is that the first three mahābhuta (महाभुत) - ākāśa (आकाश), vāyu (वायु), and agnī (अग्नी) are in possession of such qualities which are not related with the jala mahābhuta (जल महाभुत), and that is the cause we can find the attributes of these three mahābhuta (महाभुत) nearly opposite of the qualities of jala mahābhuta (जल महाभुत).

a) *vāgbhaṭṭa* (वाग्भट्ट) has described ākāśiya *dravya* (आकाशिय द्रव्य) as in possession of sukṣma (सुक्ष्म), viśada (विशद) and laghu (लघु) qualities, and suśruta (सुश्रुत) has explained that viśada (विशद) quality is a shade of rukṣa (रुक्ष) attribute. These sukṣma (सुक्ष्म), laghu (लघु), rukṣa (रुक्ष) and viśada (विशद) qualities are different in nature with the qualities of jaliya dravya (जलिय द्रव्य). *vāgbhaṭṭa* (वाग्भट्ट) has defined *āpya dravya* (आप्य द्रव्य) as holding *drava* (द्रव), śīta (शीत), *guru* (गुरु), *snigdha* (स्निग्ध), *maṁda* (मंद) and sāndra (सान्द्र) as the main qualities.

नाभसं सूक्ष्मविशदलघुशब्दगुणोल्बणम्।
अष्टांग हृदय सूत्रस्थान ९-९

b) *vāgbhaṭṭa* (वाग्भट्ट) has marked vāyaviya *dravya* (वायविय द्रव्य) as in possession of rukṣa (रुक्ष), viśada (विशद) and laghu (लघु) attributes, whereas carak has added śīta (शीत), khara (खर), sūkṣma (सूक्ष्म) qualities in the description of vāyaviya *dravya* (वायविय द्रव्य). While suśruta (सुश्रुत) has mentioned nearly the same attributes, in aṣṭāṁga saṁgraha (अष्टांग संग्रह) additional mention of vyavāyi (व्यवायि) and vikāsī (विकासी) qualities are also available. Except śīta (शीत) quality, all other qualities like rukṣa (रुक्ष), viśada (विशद), laghu (लघु), khara (खर), sukṣma (सुक्ष्म) of vāyaviya *dravya* (वायविय द्रव्य), are opposing in nature with the qualities of jaliya dravya (जलिय द्रव्य).

वायव्यं रुक्षविशदं लघु स्पर्शगुणोल्बणं । .... अष्टांग हृदय सूत्रस्थान ९-८ लघुशीतरुक्षखरविशदसूक्ष्मस्पर्श गुणबहुलानि वायव्यानि। चरक सूत्र २६-११ सूक्ष्मरुक्षखरशिशिरलघुविशदम् स्पर्शबहुलम् .... वायवियम्। सुश्रुत सूत्र ४१-५ रुक्षसूक्ष्मलघुविशदम् विकासिव्यवायीशीतखर स्पर्श गुणबहुलं वायव्यम्। अष्टांगसंग्रह सूत्रस्थान १७

c) *vāgbhaṭṭa* (वाग्भट्ट) has described āgneya *dravya* (आग्नेय द्रव्य) as in possession of rukṣa (रुक्ष), tīkṣṇa (तीक्ष्ण), uṣṇa (उष्ण), viśada (विशद) and sūkṣma (सूक्ष्म) qualities.
All other qualities like rukṣa (रुक्ष), tīkṣṇa (तीक्ष्ण), uṣṇa (उष्ण), viśada (विशद) and sūkṣma (सूक्ष्म) of āgneya *dravya* (आग्नेय द्रव्य), are

opposing in nature with the qualities of jaliya dravya (जलिय द्रव्य).

रुक्षतीक्ष्णोष्णविशदसूक्ष्मरुपगुणोल्बणम्। आग्नेयं ... ॥
अष्टांग हृदय सूत्रस्थान ९-७

As mentioned above, though *vāgbhaṭṭa* (वाग्भट्ट) has defined *āpya dravya* (आप्य द्रव्य) as holding *drava* (द्रव), śīta (शीत), *guru* (गुरु), *snigdha* (स्निग्ध), *maṁda* (मंद) and *sāndra* (सान्द्र) as the main qualities, *āyurveda darśana* (आयुर्वेद दर्शन) has specifically expressed *sneha* (स्नेह) as a special attribute of *jaliya dravya* (जलिय द्रव्य). On the other hand, *hemādri* (हेमाद्रि) has described *snigdha* (स्निग्ध) quality as one who is in possession of *kledana* (क्लेदन) power, which is elaborated by *ḍalhaṇa* (डल्हण) as moistness.

द्रवशीतगुरुस्निग्धमन्दसान्द्ररसोल्बणम्। .... अष्टांग हृदय सूत्रस्थान ९-६

स्नेहोऽपां विशेषगुणः जलवृत्तिः । आयुर्वेद दर्शन ३-२-१८-२०

यस्य क्लेदने शक्तिः स स्निग्धः। हेमाद्रि

शीतस्तिमित ....आप्यं तत्
स्नेहनह्लादनक्लेदनबन्धनविष्यंदनकरम् इति। सुश्रुत सूत्र ४१-२

क्लेदनमार्द्रभावः बन्धनं संहत्यापादनं ....डल्हण । सुश्रुत सूत्र ४१-२

All this information undoubtedly leads us towards the conclusion, that *ākāśa* (आकाश), *vāyu* (वायु), and *agnī* (अग्नी) are in possession of such qualities like, *rukṣa* (रुक्ष), *viśada* (विशद), *laghu* (लघु), *khara* (खर), *sūkṣma* (सूक्ष्म), which are nearly opposite of the qualities of *jala mahābhuta* (जल महाभुत).

विंशति गुणानाम् पुनर्विचारः

While describing the types of attributes, *vāgbhaṭṭa* (वाग्भट्ट) has classified the total 20 qualities in 2 groups, which are completely contrasting of each one. For this, *vāgbhaṭṭa* (वाग्भट्ट) has mentioned only 10 qualities namely, *guru* (गुरु), *manda* (मन्द), *hima* (हिम), *snigdha* (स्निग्ध), *ślakṣṇa* (श्लक्ष्ण), *sāndra* (सान्द्र), *mṛdu* (मृदु), *sthira* (स्थिर), *sūkṣma* (सूक्ष्म), and *viśada* (विशद). In the second line of the verse, he has used the word '*viparyayāḥ* (विपर्ययाः)' which indicates to take the 10 opposite qualities.

गुरुमन्दहिमस्निग्धश्लक्ष्णसान्द्रमृदुस्थिराः।

गुणाः ससूक्ष्मविशदा विंशतिः सविपर्ययाः॥

अष्टांग हृदय सूत्रस्थान १॥१८॥

तत्र द्रव्ये गुर्वादयो दश गुणा सविपर्यया विंशतिर्ज्ञेया ।

अरुणदत्त सर्वाङ्गसुंदर टीका

In his commentary, *āyurveda rasāyana* (आयुर्वेद रसायन), *hemādri* (हेमाद्रि) has given the 10 opposite qualities of *gururmandādi* (गुरुर्मन्दादि) attributes, namely *laghu* (लघु), *tīkṣṇa* (तीक्ष्ण), *uṣṇa* (उष्ण), *rukṣa* (रुक्ष), *khara* (खर), *drava* (द्रव), *kaṭhina* (कठिन), *cala* (चल), *sthula* (स्थुल), *picchila* (पिच्छिल).

The dictionary meaning of the word 'opposite' is: -

a. which is situated on the other or further side, when seen from a specified or contained viewpoint.
b. One more meaning is 'an entity that is totally different from or the reverse of someone or something else'.
c. One another meaning of the word 'opposite' is polar. Polar means directly opposite in

character or tendency. This adjective associated with biology, gives the meaning that relates to the poles of a cell, or organ.

Have you observed that in the first line of the original verse of *vāgbhaṭṭa* (वाग्भट्ट) reveals only 8 qualities namely, *guru* (गुरु), *manda* (मन्द), *hima* (हिम), *snigdha* (स्निग्ध), *ślakṣṇa* (श्लक्ष्ण), *sāndra* (सान्द्र), *mṛdu* (मृदु), and *sthira* (स्थिर), while remaining 2 qualities - *sūkṣma* (सूक्ष्म), and *viśada* (विशद) are explained in the next line of the verse.

Actually, these 2 *sūkṣma* (सूक्ष्म), and *viśada* (विशद) qualities belong to different characters or tendencies in comparison to first *gururmandādi* (गुरुर्मन्दादि) 8 qualities. We have already seen in earlier paragraphs that these two *sūkṣma* (सूक्ष्म), and *viśada* (विशद) qualities originate from such combination of *pāṁcabhautika* (पांचभौतिक) constitution of the dravya, where *ākāśa* (आकाश), *vāyu* (वायु), and/or *agnī* (अग्नी) have indispensable and major role.

We can also observe that the *gururmandādi* (गुरुर्मन्दादि) 8 qualities described by *vāgbhaṭṭa* (वाग्भट्ट), in the first line of this stanza, belongs to such combination of *pāṁcabhautika* (पांचभौतिक) constitution of the *dravya*, where *pṛthvi* (पृथ्वि) and *jala* (जल) *mahābhuta* (महाभुत) have crucial and main part.

तत्र गुर्वादयो दश तद्विपर्ययाश्च
लघुतीक्ष्णोष्णरूक्षखरद्रवकठिनचलस्थूलपिच्छिला दश।
हेमाद्रि आयुर्वेद रसायन टीका

Similar occurrence can be seen in the commentary of *hemādri* (हेमाद्रि), where he has demonstrated 10 opposite qualities. Among these 10 attributes, last 2 qualities are from different *pāṁcabhautika* (पांचभौतिक) origin when compared to the first 8 *laghutīkṣṇoṣṇādi*

(लघुतीक्ष्णोष्णादि) qualities. Here the last 2 qualities - *sthula* (स्थुल), *picchila* (पिच्छिल) are generated by such combination of *pāṁcabhautika* (पांचभौतिक) constitution where *pṛthvi* (पृथ्वि) and *jala* (जल) *mahābhuta* (महाभुत) have important role. On the contrary, remaining 8 qualities - *laghu* (लघु), *tīkṣṇa* (तीक्ष्ण), *uṣṇa* (उष्ण), *rukṣa* (रुक्ष), *khara* (खर), *drava* (द्रव), *kaṭhina* (कठिन), and *cala* (चल) originate from such combination of *pāṁcabhautika* (पांचभौतिक) constitution of the *dravya*, where *ākāśa* (आकाश), *vāyu* (वायु), and/or *agnī* (अग्नी) *mahābhuta* (महाभुत) have vital and most important part.

If we rearrange the last 2 qualities - *sthula* (स्थुल), and *picchila* (पिच्छिल) described by *hemādri* (हेमाद्रि), in the verse of *vāgbhaṭṭa* (वाग्भट्ट), in the place of *sūkṣma* (सूक्ष्म), and *viśada* (विशद) qualities, then all the ten attributes will have a same group of *pāṁcabhautika* (पांचभौतिक) constitutional origin in their manifestation, where *pṛthvi* (पृथ्वि) and *jala* (जल) *mahābhuta* (महाभुत) have an important role to play. The stanza should look like this -

गुरुमन्दहिमस्निग्धश्लक्ष्णसान्द्रमृदुस्थिराः।

गुणाः स स्थूलपिच्छिला विंशतिः सविपर्ययाः॥

Similarly, if we reshuffle the last 2 qualities - *sūkṣma* (सूक्ष्म), and *viśada* (विशद) described by *vāgbhaṭṭa* (वाग्भट्ट) in the last line of his stanza, with the last 2 qualities - *sthula* (स्थुल), and *picchila* (पिच्छिल) described by *hemādri* (हेमाद्रि), then all the ten attributes will have a same group of *pāṁcabhautika* (पांचभौतिक) constitutional origin in their manifestation, where *ākāśa* (आकाश), *vāyu* (वायु), and/or *agnī* (अग्नी) are major player. The line will look like this -

तत्र गुर्वादयो दश तद्विपर्ययाश्च

लघुतीक्ष्णोष्णरूक्षखरद्रवकठिनचलसूक्ष्मविशदा दश।

Now you should have followed the way, in which the direction of our discussion is moving ahead.

Very first we tried to understand a classification of amino acids, which allowed us to make two groups of amino acids, one which has a propensity to act together with water or blend or merge with water and another one which has a tendency to repel away from the water.

One should never be confused about the status of the water. Water, which we see in this Universe, is a *kārya dravya* (कार्य द्रव्य). Means it is created from the *kāraṇa dravya* (कारण द्रव्य) - *jala mahābhuta* (जल महाभुत) with the help of other four *mahābhutaḥ* (महाभुतः).

As per āyurvedika concepts, though water is in possession of *paṁca mahābhutaḥ* (पंच महाभुतः) in its fundamental constitution, actually while at the time of its manifestation, *jala mahābhuta* (जल महाभुत), followed by *pṛthvi mahābhuta* (पृथ्वि महाभुत) play important and major role. As a result of this, along with the very few attributes coming from the non-major-role- playing *ākāśa* (आकाश), *vāyu* (वायु), and *agnī* (अग्नी), water is in possession of mainly such qualities, which manifest from *jala mahābhuta* (जल महाभुत), and *pṛthvi mahābhuta* (पृथ्वि महाभुत), for example- *guru* (गुरु), *manda* (मन्द), *hima* (हिम), *snigdha* (स्निग्ध), *ślakṣṇa* (श्लक्ष्ण), *sāndra* (सान्द्र), *mṛdu* (मृदु), *sthira* (स्थिर), *sthula* (स्थुल), and *picchila* (पिच्छिल).

So when we say that hydrophobic amino acids tend to run away from the water molecules, the meaning is that there are such attributes in these hydrophobic amino acids, which are opposite in nature of the *guru*

(गुरु), *manda* (मन्द), *hima* (हिम), *snigdha* (स्निग्ध), *ślakṣṇa* (श्लक्ष्ण), *sāndra* (सान्द्र), *mṛdu* (मृदु), *sthira* (स्थिर), *sthula* (स्थुल), and *picchila* (पिच्छिल) qualities of water molecule. These are the *laghutīkṣṇoṣṇādi* (लघुतीक्ष्णोष्णादि) qualities, we have already discussed in earlier paragraphs. Having a tendency to repel from water or fail to mix with the water, these *laghutīkṣṇoṣṇādi* (लघुतीक्ष्णोष्णादि) qualities act like an enemy or antagonist of water molecule. Dictionary meaning of the word 'enemy' is 'someone who is hostile to, feels hatred towards, and opposes the interests of '. Therefore, these *laghutīkṣṇoṣṇādi* (लघुतीक्ष्णोष्णादि) qualities can be labeled as "*jalaśatru* (जलशत्रु) ".

In the matter of hydrophilic amino acids, we can say that these *gururmandādi* (गुरुर्मन्दादि) attributes which are in possession of a propensity or inclination to interact strongly with water or blend with water or merge with water, act like friend of water molecule. Therefore these *gururmandādi* (गुरुर्मन्दादि) attributes can be identified as " *jalamitra* (जलमित्र) ".

Until now we have deliberated the fundamentals of the hydrophobic amino acids and the *jalaśatru* (जलशत्रु) and *jalamitra* (जलमित्र) groups of the *viṁśatī guṇāḥ* (विंशती गुणाः) described by *vāgbhaṭṭa* (वाग्भट्ट).

आश्रयभूत ग्लायसीन : -

But in the very first sentence of this chapter, we had raised a basic query that how the *rukṣa-snigdhādi* (रुक्ष-स्निग्धादि) attributes are labeled or accredited on an entity. Means what is there, present in the dravya, by means of which the dravya is said to be in possession of a particular attribute? We have earlier stated that the *guṇāḥ* (गुणाः) or attributes need a *āśraya* (आश्रय) or shelter or support for their manifestation and it is provided by the *dravya* (द्रव्य). So let us dissect the *āśraya* (आश्रय) for better understanding of its attributes.

Glycine, alanine, methionine, phenylalanine, tryptophan, proline, valine, leucine, isoleucine, cysteine, and tyrosine come under that group, in which amino acids tend to repel from water.

Glycine is the smallest possible of all the amino acids, as it does not actually have an R group and has a hydrogen at its side chain position. Since hydrogen is non-polar, glycin becomes hydrophobic.

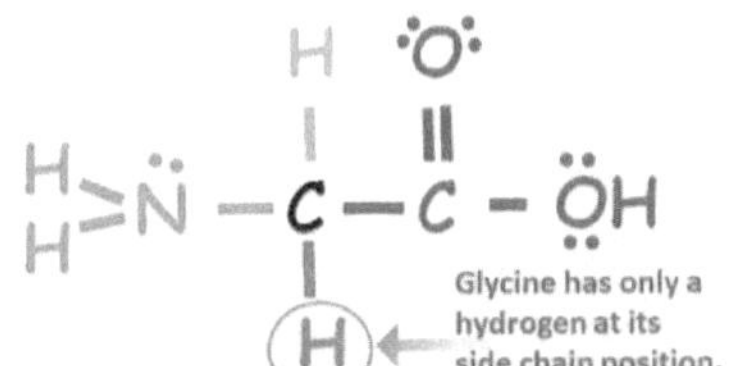
Glycine has only a hydrogen at its side chain position.

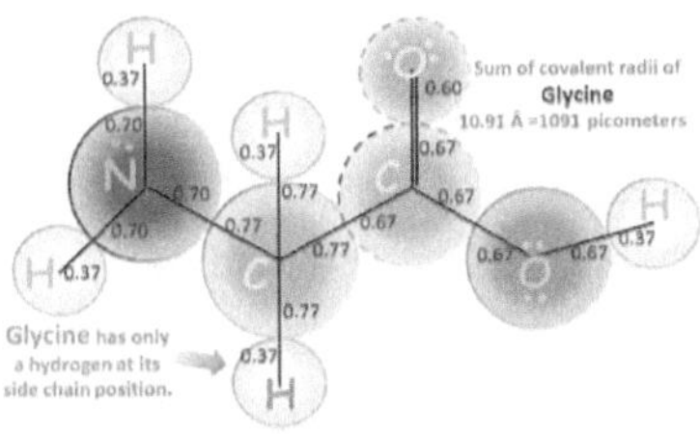
Sum of covalent radii of
Glycine
10.91 Å =1091 picometers
Glycine has only a hydrogen at its side chain position.

James M. Kovacs and others (2006) in their experiment for determining the intrinsic hydrophobicity of 23 amino acid side chains, prepared a sequence, which contains four glycine residues spread periodically throughout the sequence to ensure that the peptide has no secondary structure tendencies.

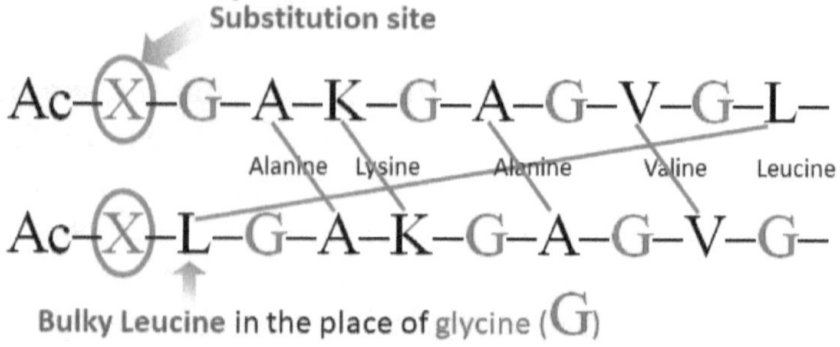

James M. Kovacs, Colin T. Mant, and Robert S. Hodges ; Determination of Intrinsic Hydrophilicity/Hydrophobicity of Amino Acid Side Chains in Peptides in the Absence of Nearest-Neighbouror Conformational Effects. Biopolymers. 2006 ; 84(3): 283–297. doi:10.1002/bip.20417

Secondly, they habituated the substitution site, adjacent to glycine residue to ensure that there is an unobstructed spin on either side of the peptide bond between the substitution site and the residue next to it.

Thirdly, to evaluate a full rotational freedom on either side of the peptide bond, they made a substitution in the sequence neighboring to a bulky Leucine residue, instead of glycine residue of the original model.

In this experiment, they found that when the substituting residue is adjacent to a glycine residue, there was complete freedom of rotation on either side of the peptide bond between substituted residue and glycine. In contrast, they also found the powerful nearest neighbor effect for all 20 amino acid substitutions adjacent to a Leucine residue.

So this experiment is very important to understand the attributes of glycine. As mentioned above, glycine has no side chain, so when in a peptide, it does not allow to advance the secondary structure. We are aware of the fact that while forming a protein, there are certain steps:

a. First, there is a sequence of amino acids, which is determined by the DNA of the gene that encodes the protein in a polypeptide chain, this level is called a primary structure.
b. Secondary structures refers to local folding form, within a polypeptide due to interactions between only atoms of the backbone. When we say only atoms of the backbone, it means R group or side chains do not participate in forming the secondary structure.
c. Afterward, a three-dimensional folding is brought together, termed as a tertiary structure, and is predominantly due to interactions between the amino acids of the side chains to form a protein. Normally up to this tertiary level, a single polypeptide chain is converted into a protein.
d. However, in the case of multi-peptide chains, which are also called as a subunit, a fourth level is necessary, in which the variable subunits interact to form a single larger protein.

Obviously the production of tertiary and/or quaternary structure is dependant on the R group or the side chain. But as we have seen that glycine has no R group, progeny from glycine is not possible for the purpose of the expansion of the protein by formation of tertiary structures. The simple meaning of this non-formation of tertiary structure is that existence of glycin becomes responsible for the restrictions of the bulkiness or enlargement of the multi-peptide chains, and ultimately inhibits the procedure of building a single larger protein. This activity is due to the presence of *laghu* (लघु) quality.

vāgbhaṭṭa (वाग्भट्ट) has described both ākāśiya *dravya* (आकाशिय द्रव्य) and *vāyaviya dravya* (वायविय द्रव्य) as in possession of *laghu* (लघु) attribute. *aruṇadatta* (अरुणदत्त) has described *laghu* (लघु) quality as opposite of *guru* (गुरु) quality, and *hemādri* (हेमाद्रि) has explained *laghu* (लघु) as such a quality, which is responsible for *laṅaghana* (लङ्घन) function.

गुरुस्तद्विपर्ययो लघु । अरुणदत्त सर्वाङ्गसुंदर टीका

लङ्घने लघु । हेमाद्रि आयुर्वेद रसायन टीका

Caraka (चरक) in his *vimānasthāna* (विमानस्थान), has described that any substance or function, which is responsible for decreasing the *guru* (गुरु) or heavy quality from the body, or increasing the lightness of the body should be recognized as *laṅaghana* (लङ्घन).

शरीरलाघवकरं द्रव्यं कर्म वा। चरक विमानस्थान

Dictionary meaning of the word 'heavy' is, "of great density; thick or substantial". One more reference is "(of food) hard to digest; excessive filling."

Both meanings are relevant to our discussion. So when we say that one of the meaning of *gurutva* (गुरुत्व) is high-density structure of the entity, its antonym *laghutva* (लघुत्व) expresses the airy, loose, with porosity, spatial arrangement resulting in low density of the structure. *Caraka* (चरक) in his *śārīrasthāna* (शारीरस्थान), has described that *laghutva* (लघुत्व) is due to the main presence of *ākāśa mahābhuta* (आकाश महाभुत) in its *pāṁcabhautika* (पांचभौतिक) constitution and furthermore it also encompasses important attributes from *vāyu mahābhuta* (वायु महाभुत) and *agnī mahābhuta* (अग्नी महाभुत) apart from *laghu* (लघु) attribute.

गुरुत्वरहितः गुणः लघुः। चरक सूत्र १।५९

लघुः आकाशगुणभूयिष्ठो वाय्वग्निगुणबहुलो गुरुविरुध्दो गुणः। चरक शारीर ६।१०

अस्य लङ्घने शक्तिः। अष्टांग हृदय सूत्र १। १५

So as discussed earlier, we can say that as per the concept described by *caraka* (चरक), glycin becomes responsible for not allowing an increase in the density of the newly forming protein with the help of *laghu* (लघु) attribute. In *āyurveda darśana* (आयुर्वेद दर्शन), very interesting definition is provided, where it says that the entity with *alpa avayavatvam* (अल्प अवयवत्वम्) or less parts similarly without or less weight should be labeled as *laghutva* (लघुत्व).

अल्प अवयवत्वम् लघुत्वम् भार रहित्वम् वा।

आयुर्वेद दर्शन अध्याय ३-२-६-७

As per *caraka vimānasthāna* (चरक विमानस्थान १।१०), the meaning of '*avayava*' means internal parts of that entity. *suśruta* (सुश्रुत) explained '*avayava*' as part of the body, while aruṇadatta (अरुणदत्त) has simplified it as the parts of the internal parts. But precise and substantial description is given by *caraka*(चरक) in his śārīrasthāna (शारीरस्थान), where he says '*avayava*' means fine, tiny parts as small as paramāṇu (परमाणु).
So when *āyurveda darśana* (आयुर्वेद दर्शन) describes 'अल्प अवयवत्वम् लघुत्वम्', the meaning is that the structure of the entity which is in possession of *laghutva* (लघुत्व) should be very tiny like paramāṇu (परमाणु).

परमाणुरुपाऽवयवांशाः। चरक शारीर ७।१७

अवयवाः घटकात्मकद्रव्यम्। चरक विमान १।१०

तदवयवाः प्रत्यङ्गानीति। सुश्रुत सूत्र ३५।१२

अवयवं अवयवं प्रति योऽवयवः तत्प्रत्यङ्गमुच्यते।

अरुणदत्त। अहृ। शा।३।१

Considering all these views of different ancient and modern texts, we can settle on some parameters for labeling an entity to be in possession of a *laghu* (लघु) attribute, the following facts should be present in it : -

1. It must have *āakāśa mahābhūta* (आकाश महाभूत), *vāyu mahābhuta* (वायु महाभुत) and/or *agnī mahābhuta* (अग्नी महाभुत) in its *pāṁcabhautika* (पांचभौतिक) constitution.
2. The structure of the entity should be very tiny and the number of internal parts of that entity should be very few. अवयवाः घटकात्मकद्रव्यम्। च वि १।१०
3. While considering the structure, the attention should reach up to the atomic level of the molecules. परमाणुरुपाऽवयवांशाः। चरक शारीर ७।१७
4. Such entities should be capable of increasing the jāṭharāgnī (जाठराग्नी)/ dhātvāgnī (धात्वाग्नी). लघून्यग्निसंधुक्षणस्वभावानि। अष्टांग संग्रह सूत्र ११
5. In the physiological transformation, the breakdown process of such an entity should be very easy and simple. शीघ्रपाकी। चरक शा। ६।१०
6. Entry of the jāṭharāgnī (जाठराग्नी)/ dhātvāgnī (धात्वाग्नी)/ pāṁcabhautikāgnī (पांचभौतिकाग्नी) in this entity should be speedy and without any problem.

7. At molecular level minimum secondary structures should be present, and lesser side chains should be in the structure.
अल्प अवयवत्वम् लघुत्वम् । आयुर्वेद दर्शन अध्याय ३-२-६-७

8. Molecular weight (in Dalton) and molecular structural length or atomic covalent radii (in Angstrom) should be minimum.
लघुत्वम् भार रहित्वम् वा। आयुर्वेद दर्शन अध्याय ३-२-६-७

9. Such entities should possess qualities which come under the "*jalaśatru* (जलशत्रु) " classification.

10. Percentage of the hydrophobic amino acids should be higher than the hydrophilic amino acids.

11. Main activities of such entities, should be the implementation of *laghutva* (लघुत्व) in the body.
अस्य लङ्घने शक्तीः। अष्टांग हृदय सूत्र १।१८

In the above paragraph, we have said that *laghutva* (लघुत्व) also incorporates important attributes from *vāyu mahābhuta* (वायु महाभुत) and *agnī mahābhuta* (अग्नी महाभुत) apart from *laghu* (लघु) attribute. In the context of discussion on glycine, here mainly we have to mention the *sūkṣma* (सूक्ष्म) quality which is present along with *laghu* (लघु) attribute in the glycine.

स्थुलविपरीतो गुणः सूक्ष्मस्रोतोगामी गुणः चरक सूत्र।१।५९

सूक्ष्ममार्गानुप्रवेशी। डल्हण २। १९-२०

अवगाहकः। सुश्रुत सूत्र ४६।५२१

विवरणे शक्तिः। हेमाद्रि १। १८

From *agni mahābhūta* (अग्नि महाभूत), *vāyu mahābhūta* (वायु महाभूत) and *āakāśa mahābhūta* (आकाश महाभूत), *any entity* acquires the ability to enter into smaller than the yocto particles (yocto means (10^{-24}) in Metric system), which is described as *sūkṣma guṇaḥ* (सूक्ष्म गुणः). *vāgbhaṭa* (वाग्भट) has mentioned that *agni mahābhūta* (अग्नि महाभूत) and *āakāśa mahābhūta* (आकाश महाभूत), both are in possession of *sūkṣma guṇa* (सूक्ष्म गुण), but in the description for attributes of *vāyu mahābhūta* (वायु महाभूत), he has not stated the *sūkṣma guṇa* (सूक्ष्म गुण) as one of its feature. Nevertheless, this *sūkṣma guṇa* (सूक्ष्म गुण) is described in the texts from *caraka* (चरक), *suśruta* (सुश्रुत), and *aṣṭāṁga* saṁgraha (अष्टांग संग्रह), as an attribute of *vāyu mahābhūta* (वायु महाभूत). On the other hand, *vāgbhaṭa* (वाग्भट) while describing *kaṭu rasa* (कटु रस), has cited its penetrating power, by mentioning it as a channel opener, by removing the obstructions and dilating the orifices, or by virtue of relaxation of smooth muscles of the capillaries. The word "*vivṛṇoti* (विवृणोति)" used in the description of *kaṭu rasa* (कटु रस), demonstrates the activities of *sūkṣma guṇa* (सूक्ष्म गुण). Hemādri (हेमाद्रि) has also cited this power as "*vivaraṇe sūkṣmaḥ* (विवरणे सूक्ष्मः)".

You may think that, what is the relevance of this passage with the amino acid glycine? In the earlier paragraphs, we have discussed how the laghutva (लघुत्व) of glycin becomes responsible for the low density of the structure of a protein. Here we will see how the *sūkṣma guṇa* (सूक्ष्म गुण) is involved in the functionality of glycine.

First, we should realize that in the movement of any substance, mainly regarding the freeness and speed of the movement, size of that entity has a major role to play. The bigger the size, we find limited movements. As the size increases, the stress-free working decreases, along with dropping the speed. In day to day life, we have observed many times that congestion reduces the speed of activity, it may be the congestion in the respiratory tract or on the high-ways. Dictionary meaning of the word 'Bulky' is "Taking up much space; large and unwieldy or too large or disorganized to function efficiently". Bulky can be compared with a word in Sanskrit as *sthula* (स्थुल), which is described by *caraka* (चरक) as which nourishes the body and which is the opposite quality of *sūkṣma guṇa* (सूक्ष्म गुण).

सूक्ष्मविपरीतो गुणः।

उपचयकरो गुणः। चरक शारीर । ६। १०

उपचयो पुष्टिः। चरक सूत्र १।८७। सुश्रुत चिकित्सा ३५।४

पुष्टिः पोषणम् तृप्तिः। चरक सूत्र २५।४०

पार्थिवद्रव्य - गुणः। सुश्रुत सूत्र ४१-४

So when we are in need of unobstructed movements along with speed, then we should forget about the entities, which are in possession of *sthula guṇa* (स्थुल गुण) and should select the substances which are full of *sūkṣma guṇa* (सूक्ष्म गुण). As we have discussed in earlier paragraphs, about the structure of glycine, now we are in opinion that glycine is definitely in possession of *laghu* (लघु) and *sūkṣma guṇa* (सूक्ष्म गुण).

The observations made by James M. Kovacs and others (2006) in their experiment mentioned at the start of this discussion, are confirming our considerations about the *pāaṁcabhautika* (पांचभौतिक) constitution of glycine.

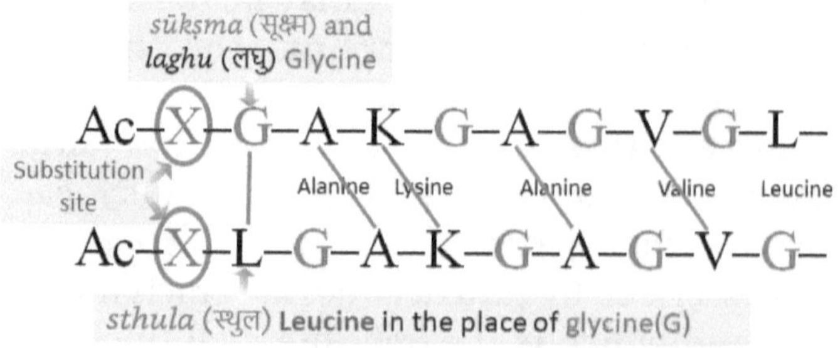

James M. Kovacs, Colin T. Mant, and Robert S. Hodges ; Determination of Intrinsic Hydrophilicity/Hydrophobicity of Amino Acid Side Chains in Peptides in the Absence of Nearest-Neighbouror Conformational Effects. Biopolymers. 2006 ; 84(3): 283–297. doi:10.1002/bip.20417

James M. Kovacs and others (2006) desired that the substitution site should in such a neighborhood that the next amino acid has a nominal or insignificant R group in terms of its size and hydrophobicity, which will permit the substituting amino acid to express its true intrinsic hydrophobicity. So they selected an amino acid-like glycine, which is in possession of a laghutva (लघुत्व) attribute, by the virtue of the presence of vāyu mahābhuta (वायु महाभुत) and *āakāśa mahābhūta* (आकाश महाभूत) in its *pāaṁcabhautika* (पांचभौतिक) constitution, and which do not facilitate any increase in the density of the newly forming protein with the help of its *laghu* (लघु) attribute. As shown in the above figure, there are four glycine residues which have no R group, and are in possession of *laghu* (लघु) attribute, spreading at regular intervals throughout the sequence, and becoming responsible for offspring inhibition and restrictions to forming of any type of alpha-helix, beta-sheet or beta-turn, resulting in lack of bulkiness or non-enlargement of the multi-peptide chains, subsequently maintaining the miniature size.
One more desire of James M. Kovacs and others (2006) was that there should be no nearest-neighbor effects. This neighboring effect mainly is due to the steric hindrance between the substituting side chain and its

nearest neighbor side chains. Steric means relating to or involving the arrangement of atoms in space.
ākāśa mahābhūta (आकाश महाभुत) facilitates the steric effect which arises from a fact that each atom within a molecule occupies a certain amount of space.
In pharmacology, steric effects determine how and at what rate a drug will interact with its target biomolecules. Steric hindrance occurs when the large size of groups within a molecule prevents chemical reactions that are observed in related molecules with smaller groups. Earlier we have discussed that larger size of groups are labeled as bulky and this bulky can be compared with *dravya* (द्रव्य) with *sthula* (स्थुल) attribute, which is manifested by the combination of *pṛthvi mahābhuta* (पृथ्वि महाभुत) and *jala mahābhuta* (जल महाभुत) in their *pāṁcabhautika* (पांचभौतिक) constitution. Thus *dravya* (द्रव्य) with *sthula* (स्थुल) attribute is one of the causes of steric hindrance.
Nearest neighbor effects can be eliminated, if we do not allow the *dravya* (द्रव्य) with *sthula* (स्थुल) attribute, to be responsible for the steric hindrance.
So the simplest way to avoid the nearest neighbor effect is to arrange the substituting side chain and its nearest neighbor side chains in such a way, that two *dravya* (द्रव्य) both having *sthula* (स्थुल) attribute, will not come together. That is the cause that, glycine which is in possession of *laghutva* (लघुत्व), is placed next to the substituting amino acid.
In short, we have seen that the amino acid, which is going to be placed next to the 'X' site (substitution site), should have a rapid turning or whirling motion, without any obstacle or interference on any side of the peptide bond between the 'X' site and the residue next to it. For the achievement of the predicted, they placed

a tiny structured amino acid, which is in possession of *laghu* (लघु) and *sūkṣma guṇa* (सूक्ष्म गुण), namely glycine, adjacent to the substitution 'X' site. One more experiment done by James M. Kovacs and others (2006) was that for evaluation of full rotational freedom, they positioned a Leucine residue at the place of glycine of the original sequence. We have seen earlier that leucine is more bulky than any hydrophobic amino acids. Actually atomic weights of glycine and leucine are 179.101 and 379.325 respectively, and atomic numbers of glycine and leucine are 92 and 196 respectively. In the scales of atomic weight and atomic numbers of all hydrophobic amino acids, glycine with its minimum numbers is placed at the lowest markings, while leucine is at the highest point. Here we can say that leucine is more bulky than glycine and in possession of *sthula* guṇa (स्थुल गुण).

So now they have placed the comparatively *sthula guṇa* (स्थुल गुण) retaining leucine, in the place of *laghu* (लघु) and *sūkṣma guṇa* (सूक्ष्म गुण) possessing glycine residue of the original sequence. They found that in this condition in the neighborhood of leucine, there was obstruction to the free movements. On the other hand, when the substituting residue was adjacent to *laghu* (लघु) and *sūkṣma guṇa* (सूक्ष्म गुण) possessing glycine residue, there was complete freedom of rotation on either side of the peptide bond between substituted residue and glycine.

Bearing in mind, all the arguments discussed in the above paragraphs, we can settle on some point of view, for labeling the glycine amino acid to be in possession of a *laghu* (लघु) and *sūkṣma* (सूक्ष्म) attribute. Glycine is in possession of *āakāśa mahābhūta* (आकाश महाभूत), and *vāyu mahābhuta* (वायु महाभुत) in its *pāṁcabhautika* (पांचभौतिक) constitution. As mentioned "**परमाणुरुपाऽवयवांशाः। चरक शारीर ७।१७**", we have seen that, in the scales of atomic weight and atomic numbers of all hydrophobic amino acids, glycine with its a.w. 179.1 and a.n. 92, is placed at the lowest markings, making the structure of the glycine very tiny. As described in " **अल्प अवयवत्वम् लघुत्वम् भार रहित्वम् वा। आयुर्वेद दर्शन अध्याय ३-२-६-७** ", glycine has no side chain, so when in a peptide, it does not allow to advance the secondary structure. As we are going to discuss the importance of presence of glycine in the structure of protein, in relation with the role of increasing the jāṭharāgnī (जाठराग्नी)/ dhātvāgnī (धात्वाग्नी) incoming paragraphs, you will find that, glycine completes the expectations mentioned in "लघून्यग्निसंधुक्षणस्वभावानि। अष्टांग संग्रह सूत्र १ ". We have also discussed that glucine possesses such qualities which come under the "*jalaśatru* (जलशत्रु) " classification.

ग्लायसीनस्य लङ्घने शक्तीः -

Subsequently, before labeling the glycine amino acid to be in possession of a *laghu* (लघु) and *sūkṣma* (सूक्ष्म) attribute, one more important criterion, is needed to be inspected. We have said earlier that main activity of such entities, should be implementation of *laghutva* (लघुत्व) in the body, as described "**अस्य लङ्घने शक्तीः। अष्टांग हृदय सूत्र १।१८**", we should able to prove the capacity of glycine to be useful as a member of food ingredients in the weight loss phenomenon. Actually, though there are many characteristics of the *laṅaghana* kārya (लङ्घन कार्य), first we will explore the *māṁsa varga* (मांस वर्ग) and *śimbīdhānya varga* (शिम्बीधान्य वर्ग), for the better understanding of the *laghu* (लघु) attribute.

1. Thorough analysis of *māṁsa varga* (मांस वर्ग): -
We are aware of the fact that *bāhya pāṁcabhautika dravyas* (बाह्य पांचभौतिक द्रव्य), which can nourish the *māṁsa dhātu* (मांस धातु), is administered in the body through the food articles. Though we have stated earlier that *māṁsa dhātu* (मांस धातु) made available from the animals, is rich in proteins, other important sources of proteins are also available from cereal grains, nuts and seeds, legumes and legumes products. As proteins are made of the amino-acid chains, it is quite clear that some of the amino acids should be in possession of the *sthira* (स्थिर), *guru* (गुरु), *snigdha* (स्निग्ध) *guṇāḥ* (गुणाः). As all the manifested dravyāḥ (द्रव्याः) are *pāṁcabhautika* (पांचभौतिक) in their constitution, we can conclude that such amino-acids, which are responsible for the nourishment of the *māṁsa dhātu* (मांस धातु) in

the body, should be manifested by those *dravyāḥ* (द्रव्याः), which are in possession of *pṛthvi mahābhuta* (पृथ्वि महाभुत) and *jala mahābhuta* (जल महाभुत) in their *pāṁcabhautika* (पांचभौतिक) constitution.

In his *annasvarupa vijñānīya* (अन्नस्वरुप विज्ञानीय) chapter, *vāgbhaṭṭa* (वाग्भट्ट) has described *māṁsa varga* (मांस वर्ग), where detail classification of the different animals and fishes are available. This classification is established on the habitat of the faunas and of course based on the characteristics of the meat available.

Initially, *vāgbhaṭṭa* (वाग्भट्ट) has classified the ten groups of wildlife and then these are again divided into 3 subgroups: - *jāṅagala* (जाङ्गल), *ānupa* (आनुप) and *sādhāraṇa* (साधारण).

1. *jāṅagala* (जाङ्गल) contains 3 sub-groups of animals.
    a. *mṛga* (मृग)
    b. *viṣkira* (विष्किर)
    c. *pratuda* (प्रतुद)

    आद्यास्त्रयो मृगविष्किरप्रतुदा - जाङ्गला।
2. *sādhāraṇa* (साधारण) contains 2 sub-groups of animals.
    a. *bileśaya* (बिलेशय)
    b. *prasaha* (प्रसह)

    मध्यौ द्वौ बिलेशयप्रसहौ - साधारणौ। आयुर्वेद रसायन –हेमाद्रि
3. *ānupa* (आनुप) contains 3 sub-groups of animals
    a. *mahāmṛga* (महामृग)
    b. *jalacara* (जलचर)
    c. *matsya* (मत्स्य)

    अन्त्यास्त्रयो महामृगजलचरमत्स्या - आनूपा।

Furthermore, *vāgbhaṭṭa* (वाग्भट्ट) has explained that the first group of *jāṅagala* (जाङ्गल) wildlife, is predominantly in possession of *laghu* (लघु) attribute, as the habitat, *jāṅagala deśa* (जाङ्गल देश) is under the umbrella of vāta doṣa (वात दोष).

आद्यान्त्या जाङ्गलानूपा मध्यौ साधारणौ स्मृतौ ।

तत्र बद्धमलाः शीता लघवो जाङ्गला हिताः॥

अष्टांग हृदय सूत्र ६।५५

जाङ्गलं वातभूयिष्ठमनूपं तु कफोल्बणम्। अष्टांग हृदय सूत्र १।२३

In his *vimānasthāna* (विमानस्थान), *caraka* (चरक) has described the *jāṅagala deśa* (जाङ्गल देश) as noticeable for its hot and excess current of air, high temperature and with scarcity of water. *Suśruta* (सुश्रुत) has additionally expressed the presence of *ākāśa mahābhūta* (आकाश महाभुत) in the description of *jāṅagala deśa* (जाङ्गल देश).

जाङ्गल देशाः अल्पोदकद्रुमाः प्रवातः प्रचुर आतपः।
चरक विमान ३।४७

आकाशसमः प्र विरलाल्प कण्टकिवृक्ष.............।
सुश्रुत सूत्र ३५।४२

In *aṣṭāṁga saṁgraha* (अष्टांग संग्रह) the attributes of the wildlife residing in an indeterminate geographic area of *jāṅagala deśa* (जाङ्गल देश) are explained in depth. The flesh of the of *jāṅagalaprāṇi* (जाङ्गलप्राणि) is recommended as the best among all types of meats and is said to be very beneficial or advantageous for the health. Though it is in possession of high nutrient values, the specialty of it is that it is very easy to digest and produces comparatively less anabolic effects in the body, which means it does not become responsible for properly promoting metabolic activities which are concerned with the biosynthesis of complex molecules.

जाङ्गलप्राणिनां मांसम् तद्गुणाः - रसे कषायमधुरं, शीतवीर्यं, लघु, विशदं, रुच्यं............मांसानां च उत्तमम् । जाङ्गल मांसम् हितकारकम्। अष्टांग संग्रह सूत्र ७-८७-८८

On the contrary, in his *sūtrasthāna* (सूत्र स्थान), *caraka* (चरक) has mentioned that the flesh of the wildlife belonging to the *ānupa deśa* (आनुप देश) is mainly in possession of *madhura rasa* (मधुर रस), *guru* (गुरु) and *snigdha* (स्निग्ध) attributes, and is responsible for increase in strength, and vigor. But he has also revealed that this flesh of animals belonging to *ānupa deśa* (आनुप देश) is only beneficial to these persons, who are regularly engaged in doing exercise and in possession of strong digestive power (दीप्ताग्नि).

आनूपदेशचारि प्राणिजं मांसम् तद् गुरु उष्णस्निग्धमधुरं बलोपचयवर्धनं वृष्यं वातहरं कफपित्तवर्धनं व्यायामदीप्ताग्निभ्यो हितम्। चरक सूत्र २७।५७

The flesh belonging to the *ānupa deśa* (आनुप देश) is said to be responsible for accumulation of *kaphavargīya dravya* (कफवर्गीय द्रव्य) in the body, due to the excessive presence of *jala mahābhuta* (जल महाभुत) in the environment.

Suśruta (सुश्रुत) has clearly stated that "*kaphavātaroga bhūyiṣṭhaśc ānūpaḥ*" means you can found more percentage of the diseases originating from the *kapha doṣa* (कफ दोष) and *vāta doṣa* (वात दोष).

अनूपसंश्रयादानूपः। चरक सूत्र २७।५४

जलभूयिष्ठो देशः - तत्र बहूदक................कफवातरोग भूयिष्ठश्चानूपः। सुश्रुत सूत्र ३५।४२

आनूपः अहितदेशानां श्रेष्ठः। चरक सूत्र २५।४०

At the start of our discussion, by testifying that "*māṁsaṁ bṛhaṇīyānāṁ śreṣṭham*" (**मांसं बृंहणीयानां श्रेष्ठम् । चरक सूत्रस्थान २५-४०**), we have self-proclaimed that, the ingested flesh material is going to be utilized for the quantitative and qualitative supplementation of muscle tissues in the body.

बृंहणम् पृथिव्यम्बुगुणभुयिष्टम्। सुश्रुत सूत्रस्थान ४१-६

As per the ayurvedic concepts, the *pāaṁcabhautika* (पांचभौतिक) constitution of flesh, mainly its *pārthiva* (पार्थिव) and *jaliya* (जलिय) parts, nurtures the *māṁsa dhātu* (मांस धातु) in the body.

लङ्घने देशमहात्म्यम् -

Now one can ask, if meat or flesh, being the result of a combination of *pārthiva* (पार्थिव) and *jaliya* (जलिय) *pāaṁcabhautika* (पांचभौतिक) constitution, becomes responsible for the *bṛṁhaṇa karma* (बृंहण कर्म) in the body, then why there is a difference in the functionality of the wildlife - flesh belongings to *jāṅagala deśa* (जाङ्गल देश) and *ānupa deśa* (आनुप देश)?

The answer is hidden in the different percentage of *pārthiva* (पार्थिव) and *jaliya* (जलिय) participation in the combination of the *pāaṁcabhautika* (पांचभौतिक) constitution of a particular type of wildlife flesh. For example, we have already discussed in the description of the *jāṅagala deśa* (जाङ्गल देश) that there is a comparatively high temperate climate. Excess current of air is present and most importantly there is always scarcity of water. The period of rainy season is very small or many times seldomly occurring, resulting in a prolonged or chronic shortage of water. All these express a particular phenomenon, in which *vāyu* mahābhuta (वायु महाभुत), and *agni* mahābhuta (अग्नि महाभुत) are present in excess percentage in the environment. We have also seen that *suśruta* (सुश्रुत) has additionally expressed the presence of *ākāśa mahābhūta* (आकाश महाभुत) in the description of *jāṅagala deśa* (जाङ्गल देश).

Second thing is that the surplus percentage is variable and unpredictable in different part of the *jāṅagala deśa* (जाङ्गल देश). For example, if we look at the state of mahārāṣṭra (महाराष्ट्र), we find that there are 2 main areas - vidarbha (विदर्भ) and marāṭhavāḍā (मराठवाडा), where we find the above-discussed climate and scarcity of water on a regular basis. Actually some districts like candrapūra (चन्द्रपूर) and bhaṁḍārā (भंडारा) have a heavy rainfall in the rainy season and these districts are recognized for propagation of paddy in large proportion. But these same districts are having very hot climate and water scarcity for the remaining 8 months. In the case of marāṭhavāḍā (मराठवाडा), we can observe that the areas near the coast of godāvarī (गोदावरी) and the districts without any big river, like usmānābāda (उस्मानाबाद), lātūra (लातूर), and bīḍa (बीड) have diverse hotness in the climate and the dissimilar rainfall, making them apart from each other in the context of *jāṅagala deśa* (जाङ्गल देश).

So these permutations and combinations or diversities in the sunshine, wind, and rainfall, manifest countless shades in the actualization of the *jāṅagala deśa* (जाङ्गल देश). As here we have mentioned a very important word "shades", for better understanding, we should look at a common example. We know that there are only 7 wavelengths within the visible spectrum of light, that each correspond to the Red, Orange, Yellow, Green, Blue, Indigo and Violet colors, but the overall sum of colors we can see is nearly around 10 million. On the other hand, a computer can exhibit almost 16.8 million colors to generate a full-color picture. Though approximately there may be an infinite number of colors, one should not forget that these numbers are dependant on the fact that, how the person can discriminate between different intensities coming from each type of photoreceptor.

So the concept behind these shades, coming from different intensities, are the base of different attributes manifesting in the flesh of wildlife of *jāṅagala deśa* (जाङ्गल देश) and *ānupa deśa* (आनुप देश).

In *jāṅagala deśa* (जाङ्गल देश) the above-discussed variations of the temperature, wind and rainfall are the result of different intensities in the combinations of *ākāśa, vāyu* and *agni mahābhūta* (आकाश, वायु, अग्नि महाभुत) in the *pāaṁcabhautika* (पांचभौतिक) constitution of a *jāṅagalaI* (जाङ्गल) type of wildlife flesh.

On the other hand, in *ānupa deśa* (आनुप देश), as discussed earlier, the excessive presence of different concentrations of *pṛthvi* (पृथ्वि) and *jala* (जल) *mahābhūta* (महाभूत) in the *pāaṁcabhautika* (पांचभौतिक) constitution of a *ānupa* (आनुप) type of wildlife flesh, causes discrepancies of the temperature, wind and rainfall in the environment.

So this is the reason that, though the same entity labeled as meat or flesh, formed by the result of a combination of *pārthiva* (पार्थिव) and *jaliya* (जलिय) *pāaṁcabhautika* (पांचभौतिक) constitution, is considered as useful for the *bṛṁhaṇa karma* (बृंहण कर्म) in the body, there is a immense dissimilarity in the functionality of the wildlife - flesh belonging to *jāṅagala deśa* (जाङ्गल देश) and *ānupa deśa* (आनुप देश).

We have mentioned here about the different intensities or concentrations of an entity. This is not a new concept. In the ancient texts, there are many references about this. For example, *vāgbhaṭṭa* (वाग्भट्ट) has explained in the description of rasa (रस) that, madhura rasa (मधुर रस), amla rasa (अम्ल रस), *lavaṇa* rasa (लवण रस), *katu rasa* (कटु रस), *tikta rasa* (तिक्त रस), and *kaṣāya rasa* (कषाय रस), these six types of rasa (रस) can be re-classified in 63 categories by mixing them with each other. He has labeled this as a sthūla (स्थूल) or gross classification.

इदानी रसानामाधारद्वारेण सयोगान् कल्पना च विभजन्नाह -

संयोगाः सप्तपञ्चाशत्कल्पना तु त्रिषष्टिधा।

रसानां यौगिकत्वेन यथास्थूलं विभज्यते॥ अष्टांग हृदय सूत्र १०। ३९

At the end of this chapter, *vāgbhaṭṭa* (वाग्भट्ट) reveals that after considering the *rasa* (रस) and *anurasa* (अनुरस), with their *tāratamyaparikalpana* (तारतम्यपरिकल्पना), the numbers of the *rasa* (रस) will be infinite or countless. Here *tāratamya* (तारतम्य) means using superlative and comparative adjectives. In English, superlative adjectives are in practice, to define an entity which is at the superior or inferior limit of a value, while comparative adjectives are used to equate dissimilarities between the two objects. In his commentary, aruṇadatta (अरुणदत्त) has given an example, stating that "This is madhura (मधुर) or sweet, this is madhuratara (मधुरतर) or sweeter, this is madhuratama (मधुरतम) sweetest".

ते रसानुरसतो रसभेदास्तारतम्यपरिकल्पनया च।

सम्भवन्ति गणना समतीतादोषभेषजवशादुपयोज्याः॥ अष्टांग हृदय सूत्र १०। ४४

तथा तारतम्यपरिकल्पनया - अय मधुरोऽय मधुरतरोऽय मधुरतम इत्येवरुपा या तया च गणना

समतीता - सङ्ख्यामतिक्रान्ता सम्भवन्ति। अरुणदत्त

Thus here, we have tried to establish, why there is a difference in the functionality of the wildlife - flesh belongings to *jāṅagala deśa* (जाङ्गल देश) and *ānupa deśa* (आनुप देश).

We have already discussed the impact of the hot and excess current of air, high temperature and scarcity of water, on the qualities of the wildlife residing in a geographic area of *jāṅagala deśa* (जाङ्गल देश). It is well established in all the ancient texts that, the flesh of the of *jāṅagalaprāṇi* (जाङ्गलप्राणि) is the unsurpassed amidst all forms of the meats, as it is in possession of high nutrient values, along with the characteristic of being very easy to digest.

जाङ्गलप्राणिनां मांसम् तद्गुणाः-रसे कषायमधुरं, शीतवीर्यं, लघु, विशदं.....अष्टांग संग्रह सूत्र ७-८७

In *aṣṭāṁga saṁgraha* (अष्टांग संग्रह), it has been classified that *jāṅagala* (जाङ्गल) type of wild-life mainly comprises 3 sub-groups of animals, namely, *mṛga* (मृग), *viṣkira* (विष्किर) and *pratuda* (प्रतुद).

आद्यास्त्रयो मृगविष्किरप्रतुदा - जाङ्गला।

Though the wildlife from all the three types of *māṁsa varga* (मांस वर्ग), namely, *jāṅagala* (जाङ्गल), *sādhāraṇa* (साधारण) and *ānupa* (आनुप), are in possession of worthy ability for quantitative and qualitative increase of the body muscles maas, resulting in fulfilling one of the benchmark of the *bṛṁhaṇa karma* (बृंहण कर्म), which is described in the ancient texts as *māṁsaṁ bṛhaṇīyānāṁ śreṣṭham* (मांसं बृंहणीयानां श्रेष्ठम् । चरक सूत्रस्थान २५-४०), *caraka* (चरक) has defined the *jāṅagala* (जाङ्गल) wildlife meat, as best amongst all the types of wildlife meat and very much beneficial, and in possession of the *laghu* (लघु) and viśada (विशद) attributes.

न हि मांससमं किञ्चिदन्यद्देहबृहत्वकृत्।

मांसादमांसं मांसेन सम्भृतत्वाद्विशेषतः॥अष्टांग हृदय सूत्र १४।३५

मांसादाना - मांससेविना मांसम् मांसेन सम्भृतत्वात् –

पुष्टत्वात् विशेषतो देहबृहत्वकृत् इत्यनुषङ्ग। हेमाद्रि

जाङ्गलप्राणिनां मांसम् तद्गुणाः - रसे कषायमधुरं, शीतवीर्यं, लघु, विशदं, रुच्यं......मांसानां च उत्तमम् । जाङ्गल मांसम् हितकारकम्। अष्टांग संग्रह सूत्र ७-८७-८८

So here we are going to give a try to understand, what is the particular difference between the wildlife meat of *jāṅagala deśa* (जाङ्गल देश) and animals from *sādhāraṇa deśa* (साधारण देश) and *ānupa deśa* (आनुप देश). We have seen that *mṛga* (मृग), *viṣkira* (विष्किर) and *pratuda* (प्रतुद) are the 3 sub-groups of animals, which come under the umbrella of *jāṅagala* (जाङ्गल) type of wild-life, while the second important group "*ānupa* (आनुप)" contains *mahāmṛga* (महामृग), *jalacara* (जलचर)and *matsya* (मत्स्य) subgroup of wildlife.

It is very difficult to analyze or compare the wildlife meat of *jāṅagala deśa* (जाङ्गल देश) and animals from *sādhāraṇa deśa* (साधारण देश) and *ānupa deśa* (आनुप देश), because majority of the animals described in these groups are not easily available. Many of these animals are not a part of daily food nowadays, as there is a law of prohibition on the hunting, selling, eating of these animals in India. Though adequate information about all the animals from each group, is not readily obtainable today, it is possible that we can evaluate the accessible facts and figures, to arrive at some conclusion.

The informal way to fathom the details of any *dravya* (द्रव्य), is to look in its *pāṁcabhautika* (पांचभौतिक) constitution. First, as a start, we will compare the structures of some of the animals from *mṛga* (मृग) sub-group of *jāṅagala deśa* (जाङ्गल देश) with the animals of *mahāmṛga* (महामृग) sub-group of *ānupa deśa* (आनुप देश) in relation with their *pāṁcabhautika* (पांचभौतिक) constitution.

	mṛga (मृग) of *jāṅagala deśa* (जाङ्गल देश)		*mahāmṛga* (महामृग) of *ānupa deśa* (आनुप देश)	
1.	Deer raw	मृगमांस	Beef –Raan gaaya	वनगव अपाचितः
2.	Deer shoulder roasted	हरिणपक्ष भर्जितः	Boar wild cooked	वराह पाचितः
3.	Deer cooked roasted	हरिण भृष्टः	Veal-trimmed raw	गोवत्स अपाचितः
4.	Rabbit wild raw	वन शशः	Buffalo raw	महिष अपाचितः
5.	Rabbit domestic raw	ग्राम्य शशः	Pork ground raw	ग्राम्य वराह
6.	Rabbit domestic roasted	शशः भर्जितः	Beef brisket	वन्य धेनु
7.	Rabbit wild cooked	शशः पाचितः	Buffalo roasted	महिष भृष्टः
8.	Rabbit domestic stewed	शशः क्वाथितः	Bison Redaa	वृषभ अपाचितः
9.	Deer ground raw venison	सारंग	Bison Gava	वन्य गौर

Table showing selected animals for comparison

Previously, we have tried to comprehend a classification of amino acids, which permitted us to create two groups of amino acids such as: -

i) One which has a susceptibility to act together with water or blend or merge with water. In his *sūtrasthāna* (सूत्र स्थान), *caraka* (चरक) has revealed that the meat of the wildlife, coming from the *mahāmṛga* (महामृग), *jalacara* (जलचर) and *matsya* (मत्स्य) sub-groups of *ānupa deśa* (आनुप देश) is essentially in custody of *madhura rasa* (मधुर रस), *gururmandsnigdhādi* (गुरुर्मन्दस्निग्धादि) qualities.

आनूपदेशचारि प्राणिजं मांसम् तद् गुरु उष्णस्निग्धमधुरं ....। चरक सूत्र २७।५७

In the same way, the meat of the wildlife coming from the *mahāmṛga* (महामृग), *jalacara* (जलचर) and *matsya* (मत्स्य), is rich in similar kinds of proteins, that are made from the combination of such amino acids, which are in possession of qualities classified under the label of "*jalamitra* (जलमित्र) ". Because these *gururmandsnigdhādi* (गुरुर्मन्दस्निग्धादि) attributes which are in possession of a propensity to interact strongly with water or merge with water, act like friend of water molecule. So we have decided that these *gururmandādi* (गुरुर्मन्दादि) attributes should be identified as "*jalamitra* (जलमित्र) ". Primarily, serine, threonine, asparagine, glutamine, aspartic acid, glutamic acid, lysine, arginine, and histidine come under this group, in which amino acids blend with water.

ii) And another one which has a propensity to drive back away from the water. We already have agreed with the fact that this tendency to repel from water or fail to mix with the water, comes due to the presence of *laghutīkṣṇoṣṇādi* (लघुतीक्ष्णोष्णादि) qualities, which act like an enemy or antagonist of water molecule. As these *laghutīkṣṇoṣṇādi* (लघुतीक्ष्णोष्णादि) qualities feel hatred towards the water molecules, we have labeled these *laghutīkṣṇoṣṇādi* (लघुतीक्ष्णोष्णादि) qualities as "*jalaśatru* (जलशत्रु) ". All the ancient texts have denoted that *ākāśa mahābhuta* (आकाश महाभुत), *vāyu mahābhuta* (वायु महाभुत), and *agnī mahābhuta* (अग्नी महाभुत) are in possession of such qualities like, *rukṣa* (रुक्ष), *viśada* (विशद), *laghu* (लघु), *khara* (खर), *sūkṣma* (सूक्ष्म), which are nearly opposite of the qualities of *jala mahābhuta* (जल महाभुत). These *jalaśatru* guṇa (जलशत्रु गुण) or *laghutīkṣṇoṣṇādi* (लघुतीक्ष्णोष्णादि) qualities need a *āśraya* (आश्रय) or shelter or support for their manifestation, which is provided by the *dravya*

(द्रव्य). It is appropriately mentioned by *caraka* (चरक) that without *dravya* (द्रव्य), *guṇāḥ* (गुणाः) are unable to manifest themselves.

गुणा गुणाश्रया नोक्ताः तस्माद्रसगुणान् भिषक् ।
चरक सूत्र २६-३६

In the context of wildlife properties, *vāgbhaṭṭa* (वाग्भट्ट) has mentioned that *mṛga* (मृग) the sub-group of *jāṅagala deśa* (जाङ्गल देश) is in possession of *laghutīkṣṇoṣṇādi* (लघुतीक्ष्णोष्णादि) qualities.

जाङ्गलप्राणिनां मांसम् तद्गुणाः ....शीतवीर्यं, लघु, विशदं, रुच्यं...। अष्टांग संग्रह सूत्र ७-८७

As the meat of these *mṛga* (मृग) is abundant in the resembling types of proteins, that are prepared from the combination of those amino acids, which are in possession of qualities classified under the label of *jalaśatru* guṇa (जलशत्रु गुण). Mainly, glycine, alanine, methionine, phenylalanine, tryptophan, proline, valine, leucine, isoleucine, cysteine, and tyrosine come under this group, in which these amino acids tend to repel from water.

मांसवर्गीय ग्लायसीन तथा लघुत्वम् -

To understand the *laghutva* (लघुत्व) of the flesh of the wild-life belonging to the *mṛga* (मृग) group, in association with the meat of the animals of *mahāmṛga* (महामृग) group, we must compare the amino acids of the meat belonging to both groups. In the following discussion, we will observe different amino acids belonging to the *jalaśatru* (जलशत्रु) group, discussed earlier.

For the discussion, we have selected nine types of flesh of the animals, which belong to the group of *mṛga* (मृग) such as, Deer raw (अपक्क मृगमांस), Deer shoulder roasted Venison (हरिणपक्ष भर्जितः), Deer cooked roasted (हरिण भृष्टः), Deer Ground Raw Venison (अपक्क सारंग), Rabbit wild raw (अपक्क वन शशः), Rabbit domestic raw (अपक्क ग्राम्य शशः), Rabbit wild cooked (वन शशः पाचितः), Rabbit domestic stewed (ग्राम्य शशः क्वथितः), and Rabbit domestic roasted (ग्राम्य शशः भर्जितः).

For the purpose of comparison, we have selected nine types of flesh of the animals, which belong to the *mahāmṛga* (महामृग) group, such as, Beef raw (वनगव अपाचितः), Bison *Gava* (वन्य धेनु), Bison- *Reda* (वन्य गौर), Boar wild cooked (वन्य वराह पाचितः भृष्टः), Beef brisket (गोवत्स पाचितः), Buffalo roasted (महिष पाचितः भृष्टः), Veal raw (गोवत्स अपाचितः), Boar wild raw (वन्य वराह अपाचितः), Pork domestic raw (ग्राम्य वराह अपाचितः), Buffalo raw (महिष अपाचितः), Pork cured ham patties (वराह पाचितः), Veal- trimmed cuts cooked (गोवत्स त्वष्ट्र पाचितः भृष्टः), Veal-trimmed cuts raw (गोवत्स त्वष्ट्र अपाचितः).

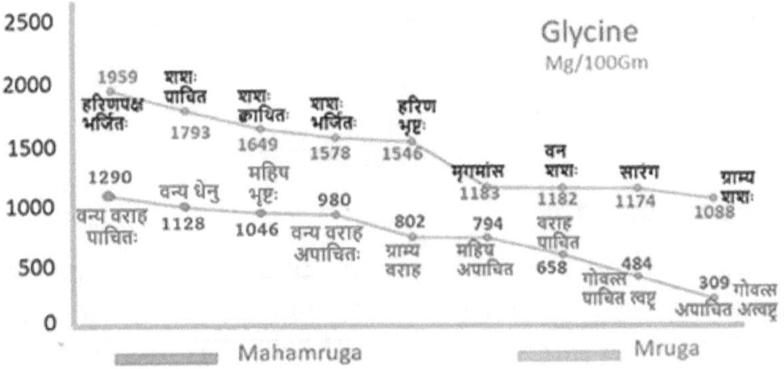

Table showing the high levels of Glycine in *mṛga* (मृग) group.

If we observe the above table, we can find that, the brown line showing the glycine quantity of wild-life meat, belonging to the *mṛga* (मृग) group, is undoubtedly with higher values than the values of blue line, which are indicative of the glycine qualities of wildlife meat belonging to *mahāmṛga* (महामृग) group. We have already discussed that, the presence of glycine in a molecule demonstrates the presence of *vāyu mahābhūta* (वायु महाभूत), *agni mahābhūta* (अग्नि महाभूत) and *āakāśa mahābhūta* (आकाश महाभूत), which are essential components of the *laghu* (लघु) attribute. As we have already seen the influence of *jāṅagalav deśa* (जाङ्गल देश), *ānupa deśa* (आनुप देश) and *sādhāraṇa deśa* (साधारण देश) in the manifestation of attributes, particularly of *laghu* (लघु) quality, similarly, *kāla* (काल) or season of the life/year, has also a specific role while determining the attributes of any entity. *vāgbhaṭṭa* (वाग्भट्ट) has classified this impact of *kāla* (काल) in three groups, and hemādri (हेमाद्रि) in his commentary has revealed the fact that *ādirbālyāvasthā kaphasya* (आदिर्बाल्यावस्था कफस्य) means in the childhood *kapha* (कफ) is more prominent than other *doṣa* (दोष).

वयोहोरात्रिभुक्तानां तेऽन्तमध्यादिगाः क्रमात्। अष्टांग हृदय सूत्र।१।८

दोषकालानाह-वय इति। तत्र वय शरीर परिणाम...............आदिर्बाल्यावस्था कफस्य।

आयुर्वेद रसायन- हेमाद्रि।

As *kapha doṣa* (कफ दोष) is in possession of *pṛthvi mahābhuta* (पृथ्वि महाभुत) and *jala mahābhuta* (जल महाभुत) in their *pāṁcabhautika* (पांचभौतिक) constitution, obviously, all the living entities on the planet are having *pṛthvi mahābhuta* (पृथ्वि महाभुत) and *jala mahābhuta* (जल महाभुत) predominantly in their *pāṁcabhautika* (पांचभौतिक) constitution at the time of their childhood, because as revealed by hemādri (हेमाद्रि), *bālyāvasthā* (बाल्यावस्था) is mostly governed by *kapha doṣa* (कफ दोष).

When we say that at *bālyāvasthā* (बाल्यावस्था), there is a predominance of *pṛthvi mahābhuta* (पृथ्वि महाभुत) and *jala mahābhuta* (जल महाभुत) in the *pāṁcabhautika* (पांचभौतिक) constitution of the body, the concealed meaning is that, other three mahābhuta (महाभुत) - ākāśa (आकाश), vāyu (वायु), and agnī (अग्नी), have a subsidiary role to carry out in the *bālyāvasthā* (बाल्यावस्था), and they are powerless to demonstrate their existence in the manifestation of the attributes. This phenomenon leads to the diminished presentation of *rukṣasūkṣmādi* (रुक्षसूक्ष्मादि) qualities in the body.

The purpose of all this discussion leads us towards the conclusion that, in the *bālyāvasthā* (बाल्यावस्था), *gururmandsnigdhādi* (गुरुर्मन्दस्निग्धादि) qualities are invariably excess in quantities, and *rukṣasūkṣmādi* (रुक्षसूक्ष्मादि) qualities are in much less extents in the body. If again, we have a look at the table showing the levels of glycine displayed above, we can detect that, the last two lowest numbers (309 and 484) from the values of blue line, which are indicative of the glycin qualities of wild-life meat belonging to *mahāmṛga* (महामृग) group, are designated to the flesh of Veal. Actually, Veal denotes meat of calves and beef signifies meat of grown-up cattle. Usually after reaching 18 to 20 weeks of the age, and weight is about 200 kg, calves are slaughtered. While the natural life span is of 15-20 years, typical slaughter age of "Veal" calves is 1-24 weeks, which clearly indicates that, calves are slaughtered when they are at *bālyāvasthā* (बाल्यावस्था), and at this stage of life, *gururmandsnigdhādi* (गुरुर्मन्दस्निग्धादि) qualities in their body, are invariably dominant, in comparison to the *rukṣasūkṣmādi* (रुक्षसूक्ष्मादि) qualities in the body. So it is not astonishing that the values of glycine look relatively less in the above table. Our study has shown that in Veal flesh, the *gururmandsnigdhādi* (गुरुर्मन्दस्निग्धादि) qualities are approximately three times more than the *rukṣasūkṣmādi* (रुक्षसूक्ष्मादि) qualities present in it. The second observation about these two values of Veal flesh is that 484 is the value of cooked and trimmed Veal cuts, while 309 is the value of raw Veal flesh. This difference in these two values is due to the effect of cooking the Veal, of course, the result of the application of agnī (अग्नी).

Considering the effect of agnī (अग्नी) on the values of raw flesh, we can observe that the values of glycine increases after the application of agnī (अग्नी), as shown in the following table: -

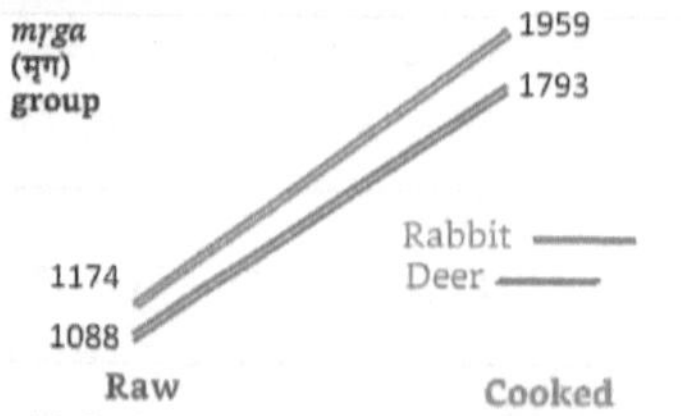

Table showing the change in Glycine values after processing

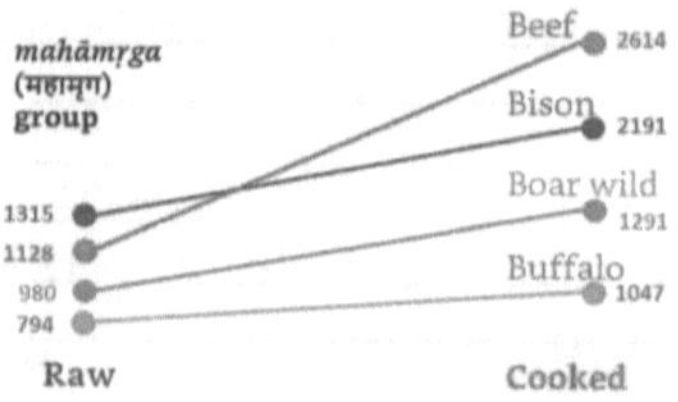

Table showing the change in Glycine values after processing

On the left side of the table, we can see the meats of two animals from *mṛga* (मृग) group, in which the glycine values of the flesh of rabbit has amplified by 66.86%, while the glycine values of the flesh of deer has increased by 64.79%.

On the right side of the table, these are the meats of some animals from *mahāmṛga* (महामृग) group, in which the glycine values of the flesh of buffalo, wild boar, bison, and beef have been shown to be increased by 31.86%, 31.73%, 66.61% and 131.73% respectively.

These variances are due to the implementation of agni in the different processings for the cooking, which we will discuss later in the description of suxma guna and agni.

We have seen that there are 3 sub-groups of *jāṅagala* (जाङ्गल) type of wild-life animals such as *mṛga* (मृग), *viṣkira* (विष्किर), and *pratuda* (प्रतुद).

आद्यास्त्रयो मृगविष्किरप्रतुदा - जाङ्गला।

In earlier pages, we have discussed in details about the flesh of *mṛga* (मृग) type of wildlife. Now we will have a look at the *viṣkira* (विष्किर), and *pratuda* (प्रतुद) types of wildlife.

लाववार्तीकवर्तीररक्तवर्त्मककुक्कुभाः।

कपिञ्जलोपचक्राख्यचकोरकुरुबाहवः॥४४

वर्तको वर्तिका चैव तित्तिरिः क्रकरः शिखि।

ताम्रचूडाख्यबकरगोनर्दगिरिवर्तिकाः॥४५

तथा शारपदेन्द्राभवरटाद्याच विष्किरा। अष्टांग हृदय सूत्र ६

In the following paragraphs, we will observe different amino acids belonging to the *jalaśatru* (जलशत्रु) group, argued earlier. For the discussion, we have selected five types of flesh of the animals, which belong to the group of *viṣkira* (विष्किर) such as, lāvā (लावा), kapota (कपोत), haṁsa (हंस), tāmracūḍa or kukkuṭa (ताम्रचूड तथा कुक्कुट), and bhāradvāja (भारद्वाज).

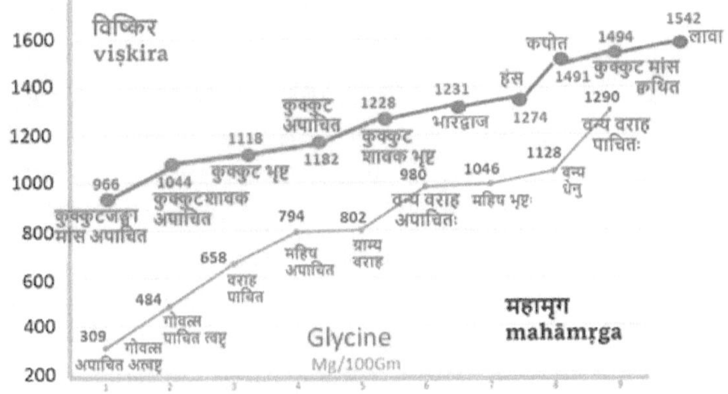

This graph shows the difference in the values of glycine from the flesh belonging to the *mahāmṛga* (महामृग) and *viṣkira* (विष्किर) wild-life animals.

In the group of *viṣkira* (विष्किर), the brown-colored names and figures are indicative of these animals which are rarely consumed nowadays. In some part of the nation, like a north-east area, till birds belonging to the *viṣkira* (विष्किर) group of wild-life are utilized as a part of daily food, but majority these brown colored names are out of the spectra of our food. But the figures and names given in pink color are a regular part of Indian food. We have specially mentioned six types of meat of chicken which are consumed in the vast proportion of mass.

kukkuṭaśāvaka (कुक्कुटशावक) or a young chicken, after processed by either roasting, grilling or barbecuing, becomes more *laghu* (लघु) in its nature due to the increase in the qualities belonging to *jalaśatru* (जलशत्रु) group of the *viṁśatī guṇāḥ* (विंशती गुणाः) described by *vāgbhaṭṭa* (वाग्भट्ट). We can observe in the above graph that the raw kukkuṭaśāvaka (कुक्कुटशावक) is in possession of the glycine value of 1044 mg/100 gm, while after cooked, the glycine value is improved to 1228 mg/100 gm. This upsurge of 184 mg/100 gm numbers in the value of glycine, reveals the intensification of *laghu* (लघु) quality in the processed kukkuṭaśāvaka (कुक्कुटशावक). In the matter of matured kukkuṭa (कुक्कुट), the raw bird gives the value of glycine as 966 mg/100 gm, while after cooked slowly in liquid, in a closed dish or pan, the meat of the kukkuṭa (कुक्कुट) turns out to be in possession of 1494 mg/100 gm of glycine, with nearly 55% rise in the glycine quality. This reveals the fact that though the birds belonging to the Phasianidae family, are basically in possession of more *laghu* (लघु) quality, in comparison with the animals of *mahāmṛga* (महामृग) group of animals, if these chicken birds are cooked slowly on low heat for a prolonged period, they become transformed having more *laghu* (लघु) quality in their possession.

Here we can perceive that, the pink and brown mixed colored line, showing the glycine quantities of wild-life meat, belonging to the *viṣkira* (विष्किर) group, is indisputably with greater values than the values of blue line, which are suggestive of the glycin qualities of wildlife meat having its place in *mahāmṛga* (महामृग) group. The average value of the glycine from the flesh of wild-life of *viṣkira* (विष्किर) group is 1256.9 mg/100 gm, while the average value of glycine of the flesh of animals belonging to *mahāmṛga* (महामृग) group is found to be 709.7 mg/100 gm only. Therefore, it evidently exhibits that, in the given sample the meat of the animals from *viṣkira* (विष्किर) group are in possession of 77.1% more glycine, than the glycine available in the meat of animals from *mahāmṛga* (महामृग) group. Now it has been already proved in previous discussion that the presence of glycine in a molecule, demonstrates the presence of *vāyu mahābhūta* (वायु महाभूत), *agni mahābhūta* (अग्नि महाभूत) and *āakāśa mahābhūta* (आकाश महाभूत), and these three *mahābhūta* (महाभूत) are truly crucial constituents of the *laghu* (लघु) attribute.

उपमेय हेतु ग्लुटामीक एसीड ज्ञानम्

As we have agreed on the fact that, the informal way to fathom the details of any *dravya* (द्रव्य), is to look in its *pāṁcabhautika* (पांचभौतिक) constitution. So now we will compare the structures of some of the food article belonging to *śimbīdhānya varga* (शिम्बीधान्य वर्ग). Actually the majority of Indian vegetarian diet come from the two major groups - śūka varga (शूक वर्ग) and *śimbīdhānya varga* (शिम्बीधान्य वर्ग), described in the *annasvarupa vijñānīya* (अन्नस्वरुप विज्ञानीय) chapter by *vāgbhaṭṭa* (वाग्भट्ट).

Before going further, we should have a small discussion, why we are considering the values of glutamic acid here. Glutamic acid is one of the important amino acids responsible for protein synthesis and synthesis of neurotransmitters like GABA. There are 3 carbons, 5 hydrogens and 2 oxygens in the side-chain glutamic acid, which has make the amino acid polar, which means it has an inclination to interact strongly with water. We have seen that water is in its possession, a high percentage of combination of *jala mahābhuta* (जल महाभुत), and *pṛthvi mahābhuta* (पृथ्वि महाभुत) mainly, in comparison with the percentage of - ākāśa (आकाश), vāyu (वायु), and agnī (अग्नी) mahābhuta (महाभुत) in its *pāṁcabhautika* (पांचभौतिक) constitution. The extreme presence of *jala mahābhuta* (जल महाभुत) in the constitution of water is marked by *caraka* (चरक) while treating the atyagni (अत्यग्नि).

गोधूमचूर्णमन्थोऽत्यग्निनाशनः। चरक चिकित्सा १५।२२६

As a result of this, water is in possession of mainly such qualities, which manifest from *jala mahābhuta* (जल महाभुत), and *pṛthvi mahābhuta* (पृथ्वि महाभुत), for example- *guru* (गुरु), *hima* (हिम), *snigdha* (स्निग्ध), *ślakṣṇa* (श्लक्ष्ण), and *mṛdu* (मृदु).

These *gururhimādi* (गुरुर्हिमादि) attributes are in possession of a propensity to blend with water or merge with water and act like a friend of water molecule. As glutamic acid has a propensity towards water, we can say that these *gururhimādi* (गुरुर्हिमादि) attributes which are labeled as "*jalamitra* (जलमित्र)" are also present in the glutamic amino acids.

When we say that *gururhimādi* (गुरुर्हिमादि) attributes can be labeled as "*jalamitra* (जलमित्र)", then we should laid some criteria for the recognition of these *gururhimādi* (गुरुर्हिमादि) attributes in an entity.

1. It must have *jala mahābhuta* (जल महाभुत), and *pṛthvi mahābhuta* (पृथ्वि महाभुत) as a major stakeholder, in its *pāṁcabhautika* (पांचभौतिक) constitution.
2. The structure of the entity should be considered broad, spacious and substantial and the number of internal parts of that entity should be plenty.
3. While considering the structure, the attention should reach up to the atomic level of the molecules. We know that, is the element bulky or insubstantial should be decided on the mean distance from the center of the nucleus to the boundary of the surrounding shells of electrons, which is also called as atomic radius of that element. This atomic radius is majored in angstrom unit. We have already observed that higher the values of distance, more bulky the

atom. Here glutamic acid has a covalent radius of 22.57 angstrom, which is 13th in the rank of 20 amino acids in view of bulkiness. Phenylalanine is the lowest in the rank with 10.61-angstrom value, while arginine is at top of the bulkiness within possession of 31 angstrom. We have seen that 'Bulky' means which takes up extra space, or which is big in size. We have discussed in earlier paragraphs that bulky can be matched with a word in Sanskrit as *sthula* (स्थुल), which is termed by *caraka* (चरक) as which sustains the body and which is the conflicting in nature of *sūkṣma guṇa* (सूक्ष्म गुण). This leads us to the conclusion that glutamic acid is in possession of *sthula* (स्थुल) attribute.

4. As such entities are bulky and with many parts, their disintegration is a major problem for the jāṭharāgnī (जाठराग्नी)/ dhātvāgnī (धात्वाग्नी). Presence of excess amount of glutamic acid becomes responsible for the increase in the *sthula* (स्थुल) quality of the entity and this thickness inhibits the deep penetration of agnī (अग्नी), while the digestion is in progress. Therefore, as a rule, the entities belonging to this group will be digested very slowly. Not only in the digestion at the abdominal level, but in the metabolic activities such as in the physiological transformation, breakdown process of such entity is with slow motion, sluggish and time-taking.

अयं चिरपाकी। सुश्रुत सूत्र ४५। १९८

5. *hemādri* (हेमाद्रि) has also described that the entity possessing *guru* (गुरु) quality has such

constitutional *mahābhuta* (महाभुत) in it, which are exactly opposite of jāṭharāgnī (जाठराग्नी)/ dhātvāgnī (धात्वाग्नी) in the body, so that such entities are responsible for the derangement of jāṭharāgnī (जाठराग्नी)/ dhātvāgnī (धात्वाग्नी) and entry of the jāṭharāgnī (जाठराग्नी)/ dhātvāgnī (धात्वाग्नी)/ pāṁcabhautikāgnī (पांचभौतिकाग्नी) in this entity is very difficult due to the presence of *guru* (गुरु)and *sthula* (स्थुल) attributes.

अयमग्निगुणविपरीतत्वादग्निमान्द्यकरः।
हेमाद्री अष्टांग संग्रह सूत्र। ११

6. Such entities should possess qualities which come under the "*jalamitra* (जलमित्र) " classification.
7. Percentage of the hydrophilic amino acids should be higher than the hydrophobic amino acids.
8. Main activities of such entities, should be the implementation of bṛṁhaṇa (बृंहण) in the body.

गुणतः पाकतश्च यद्गौरवकरं तद्द्रव्यम्। सुश्रुत उत्तरस्थान ५०।३०

तस्य कर्माणि अग्निसादोपलेपबलोपचयतर्पणबृंहणानि। सुश्रुत सूत्र ४६। ५१७

So this was a background of our determination, to select the glutamic acid for the comparison with glycine. Now we will first discuss the two most important food ingredients of Indian diet, first one is godhūmaḥ (गोधूमः) and second one is Sorghum(देवधान्य), or generally known as jvārī (ज्वारी).

Though godhūmaḥ (गोधूमः) comes under the classification of śūka varga (शूक वर्ग), the percentage of its consumption in Indian diet is too high and the benefits of godhūmaḥ (गोधूमः) described by *vāgbhaṭṭa* (वाग्भट्ट), are of so important, that we have decided to consider it for comparison with the food articles belonging to *śimbīdhānya varga* (शिम्बीधान्य वर्ग).

Different types of wheat or godhūmaḥ (गोधूमः) are available in the market, but godhūmaḥ (गोधूमः) can be classified mainly on its harvesting time of the year. Throughout the wheat harvesting countries, wheat is cultivated twice in year, and so named as wheat hard red winter and wheat hard red spring. As earlier discussed, there are *jalaśatru* (जलशत्रु) and *jalamitra* (जलमित्र) groups of the *viṁśatī guṇāḥ* (विंशती गुणाः), which are described by *vāgbhaṭṭa* (वाग्भट्ट) and amongst these two groups, *gururmandādi* (गुरुर्मन्दादि) attributes which can be identified as " *jalamitra* (जलमित्र) ", are manifested from *jala mahābhuta* (जल महाभुत), and *pṛthvi mahābhuta* (पृथ्वि महाभुत), for example- *guru* (गुरु), *manda* (मन्द), *hima* (हिम), *snigdha* (स्निग्ध), *ślakṣṇa* (श्लक्ष्ण), *sāndra* (सान्द्र), *mṛdu* (मृदु), *sthira* (स्थिर), *sthula* (स्थुल), and *picchila* (पिच्छिल).

वृष्यः शीतो गुरुः स्निग्धो जीवनो वातपित्तहा।

सन्धानकारी मधुरो गोधूमः स्थैर्यकृत्सरः। अष्टांग हृदय सूत्र ६ । १५

vāgbhaṭṭa (वाग्भट्ट) has described, wheat or godhūmaḥ (गोधूमः) as in possession of *guru* (गुरु), *snigdha* (स्निग्ध), and *sthira* (स्थिर) qualities. As these 3 attributes come under the classification of " *jalamitra* (जलमित्र)", obviously *laghutīkṣṇoṣṇādi* (लघुतीक्ष्णोष्णादि) attributes should be present with less quantities in wheat or godhūmaḥ (गोधूमः). We have already discussed that glycine is definitely in possession of *laghu* (लघु) and *sūkṣma guṇa* (सूक्ष्म गुण). So while studying the *laghutīkṣṇoṣṇādi* (लघुतीक्ष्णोष्णादि) attributes, we should always have an eye on the values of glycine. Even we can say that minimum the value of glycine, greater the chances of the entity being in possession of *gururmandādi*

(गुरुर्मन्दादि) attributes and higher the value of glycine, then there are chances of the entity being in possession of *laghutīkṣṇoṣṇādi* (लघुतीक्ष्णोष्णादि) attributes.

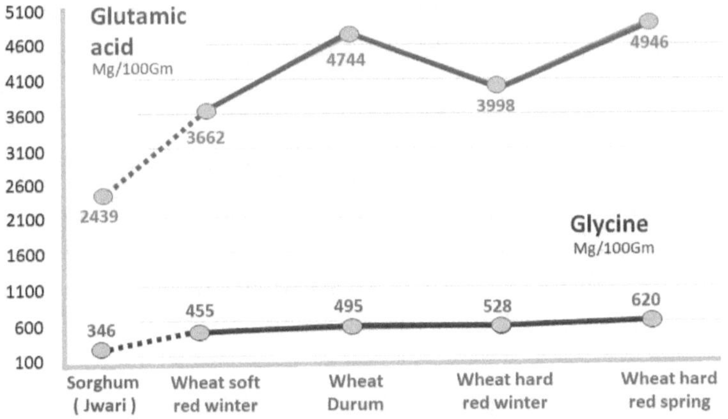

In rural India, roṭī (रोटी) or bhākarī (भाकरी) is prepared from the flour of Sorghum(देवधान्य), generally known as jvārī (ज्वारी), which is one of the principal sources of nutrition. Wheat, having a relatively higher percentage of proteins (13%), in comparison to other major cereals, is widely used in north India. In the graph presented above, we can observe that the brown-colored figure and line reveal the glutamic acid value of Sorghum(देवधान्य), which is 2439 mg/100gm, while the glutamic acid value of winter wheats (गोधूम रब्बी)- soft red, durum, hard red and hard red spring are 3662, 4744, 3998 and 4946 mg /100 gm respectively. It clearly suggests that the wheat (गोधूम) is in possession of more glutamic acid than that of Sorghum(देवधान्य).

Observing the blue line and figures in the above chart, we can determine that the glycine value of Sorghum(देवधान्य), which is 346 mg/100gm, while the glycine value of winter wheats (गोधूम रब्बी)- soft red, durum, hard red and hard red spring are 455, 495, 528, and 620 mg / 100 gm respectively.

For understanding of *laghutīkṣṇoṣṇādi* (लघुतीक्ष्णोष्णादि) attributes of the Sorghum(देवधान्य) and different types of wheat (गोधूम), if we consider the ratio of glycine and glutamic acid in Sorghum(देवधान्य), it appears to be of 1:7 (glycine 346 mg/100gm: glutamic acid 2439 mg/100gm).

If we assume that there is zero difference in between the *laghutīkṣṇoṣṇādi* (लघुतीक्ष्णोष्णादि) attributes of Sorghum(देवधान्य) and different types of wheat (गोधूम), then the values of glycine and glutamic acid should be in the ration of 1:7 as shown in the following table.

Comparison of Glycine and Glutamic acid of Sorghum and different Wheats

	Sorghum	Wheat-1	Wheat-2	Wheat-3	Wheat-4
Glycine	0346	0455	0495	0528	0620
Glutamic acid as per ratio	2439	3207	3439	3772	4370
Actual Glutamic acid	2439	3662	4744	3998	4946
Difference	0000	0455	1305	0276	0576

Base of calculation is Glycine : Glutamic acid ratio of Sorghum = 1.00 : 7.04

Winter Wheat - (गोधूम रब्बी)- 1.Soft red, 2.Durum, 3.Hard red and 4.hard red spring

But there is significant difference in the actual values as shown in the above table, which reveals that our assumption that there is zero difference in between the *laghutīkṣṇoṣṇādi* (लघुतीक्ष्णोष्णादि) attributes of Sorghum(देवधान्य) and different types of wheat (गोधूम), is rejected by this observation. In simple words, we can say that Sorghum(देवधान्य) is definitely in possession of *laghu* (लघु) and *sūkṣma guṇa* (सूक्ष्म गुण), while different types of Wheat are showing the substantial excessive existence of *gururmandādi* (गुरुर्मन्दादि) attributes in comparison to the values of Sorghum(देवधान्य). Basically one should remember that *vāgbhaṭṭa* (वाग्भट्ट) has described, wheat or godhūmaḥ (गोधूमः) as in

possession of *guru* (गुरु), *snigdha* (स्निग्ध), and *sthira* (स्थिर) qualities. Means godhūmaḥ (गोधूमः) is in possession of a high percentage of combination of *pṛthvi mahābhuta* (पृथ्वि महाभुत) and *jala mahābhuta* (जल महाभुत) mainly, in comparison with the percentage of - ākāśa (आकाश), vāyu (वायु), and agnī (अग्नी) mahābhuta (महाभुत) in its *pāṁcabhautika* (पांचभौतिक) constitution. Therefore when we say that *guru* (गुरु), *snigdha* (स्निग्ध), and *sthira* (स्थिर) attributes which come under the classification of " *jalamitra* (जलमित्र)" are present in higher quantities, obviously *laghutīkṣṇoṣṇādi* (लघुतीक्ष्णोष्णादि) attributes should be present with less quantities in wheat or godhūmaḥ (गोधूमः).
One more thing to consider is the presence of ṛtu (ऋतु) or kāla (काल), which shows a significant impact on the harvesting of food crops. As described by *caraka* (चरक), in the spring, summer and rainy season, the *gururmandādi* (गुरुर्मन्दादि) attributes belonging to the group, identified as " *jalamitra* (जलमित्र) ", are sponged up or sucked by the rays of Sun, resulting in enhancing the *laghutīkṣṇoṣṇādi* (लघुतीक्ष्णोष्णादि) qualities which are labelled as "*jalaśatru* (जलशत्रु) ". We have seen that glycine is such an amino acid, that gives *āśraya* (आश्रय) or support for the manifestation of *laghutīkṣṇoṣṇādi* (लघुतीक्ष्णोष्णादि) attributes. Therefore we should be able to see a considerable rise in the glycine value of these crops which are harvested in the ādānakāla (आदानकाल)(खरीप).

आददाति क्षपयति पृथिव्याः सौम्यांशं प्राणिनां च बलमित्यादानम्। चरक सूत्र ६।४

वसंतग्रीष्मवर्षास्त्रय ऋतवः आदानकालः। चरक चिकित्सा ३।४७

Quite the reverse, at the time of description of the visargakālaḥ (विसर्गकालः), suśruta (सुश्रुत) has revealed that the meaning of visarga (विसर्ग) is that particular activity which creates or brings something into existence. He also specifies that visarga (विसर्ग) means production of strength in the body. As described by *caraka* (चरक), in the autumn or fall, winter and frost season, the *gururmandādi* (गुरुर्मन्दादि) attributes belonging to the group, identified as " *jalamitra* (जलमित्र) ", are enhanced or enriched, not only in the body but also in the environment. He prescribes this period as the proper time to regenerate the strength of human being.

विसर्गम् सर्जनम् बलादिजननम्। सुश्रुत सूत्र २१।८

विसृजति जनयति सोमांशं प्राणिनां च बलमिति विसर्गः। चरक सूत्र।६।४

शरद् हेमन्तशिशिरास्त्रय ऋतवो विसर्गकालः। चरक चिकित्सा ३।४५

We have seen that glutamic acid is such an amino acid, that gives *āśraya* (आश्रय) or support for the manifestation of *gururmandādi* (गुरुर्मन्दादि) attributes. Therefore we should be able to see a considerable rise in the glutamic acid value of these crops which are harvested in the visargakālaḥ (विसर्गकालः). Comparing to the glycine and glutamic values of Sorghum(देवधान्य) and all types of godhūmaḥ (गोधूमः), we can conclude that Sorghum(देवधान्य) has less nutritional values than all types of wheat.

Until now, we have discussed two important entities - godhūmaḥ (गोधूमः) and Sorghum(देवधान्य), which come under the classification of śūka varga (शूक वर्ग), and their percentage of consumption in Indian diet is too high to neglect.

शिम्बीवर्गीय ग्लायसीन तथा लघुत्वम्

Now we will move further, towards our study of next group of the food articles, which belong to *śimbīdhānya varga* (शिम्बीधान्य वर्ग) as per description available in the aṣṭāṁga hṛdaya (अष्टांग हृदय).

मुद्गाढकीमसूरादि शिम्बिधान्यं विबन्धकृत्।

कषायं स्वादु सङ्ग्राहि कटुपाकं हिमं लघु॥अष्टांग हृदय सूत्र ६।१७

यत् शिम्बीधान्य तद्विबन्धकृत्। केषा विबन्ध करोति सामर्थ्यात् स्रोतसाम् न तु पुरिषादीनाम्।.....मसूरादीत्यत्राऽऽदिशब्देन मकुष्ठचणकादीना ग्रहणम्।अरुणदत्त

For the convenience, we have selected some of the *śimbīdhānya* (शिम्बीधान्य) which are extensively used in the Indian diet, like green peas (कलाय), french beans (श्रावण घेवडा), adzuki beans (लाल चवळी), pigeon peas (तूर डाल), *caṇaka* (चणक), black beans (घेवडा), *makuṣṭha* (मकुष्ठ), lima beans (पावटा), mung beans (मूग), rajmaasha (राजमाष), *masura* (मसुर) and *māṣa* (माष), and for the comparison we have considered the Sorghum (देवधान्य) and all types of godhūmaḥ (गोधूमः), because we have already revealed the fact that Sorghum(देवधान्य) has less nutritional values than all types of wheat.

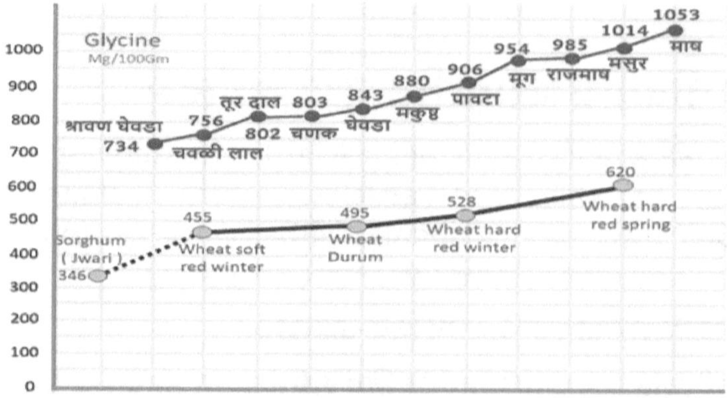

When we try to understand the blue line and figures in the above chart, we can see that the glycine value of Sorghum(देवधान्य), which is 346 mg/100gm, and the glycine values of winter wheats (गोधूम रब्बी)- soft red, durum, hard red winter and hard red spring are 455, 495, 528, and 620 mg /100 gm respectively. A glance at the graph reveals that these glycine values are definitely in lower quantity than the glycine values of the selected *śimbīdhānya* (शिम्बीधान्य), which are presented in the graph by purple line and figures. Highest value of the glycine from the Sorghum(देवधान्य) and wheat (गोधूम रब्बी) group is lower than 100mg/100gm of the glycine value of the lowest value holder of the selected *śimbīdhānya* (शिम्बीधान्य) group.

For instance, we have already discussed that the presence of higher values of glycine in any food article is indicative of the functionality of *laghu* (लघु) attribute. If considering the above graph in a hurry, one may conclude that as the highest value of the glycine amino acid displayed in the graph is 1053 mg/100 gm of *mungo beans* (माष), then the *māṣa* (माष) is in possession of more propensity towards the *laghutva* (लघुत्व) functionality. But this conclusion is based on wrong assumptions, because the graph is showing only the glycine amino acid value, which is related to *laghu* (लघु) attribute. Before labeling any attribute or *guṇaḥ* (गुणः), we know that the *guṇaḥ* (गुणः) or attribute needs a *āśraya* (आश्रय) or shelter or support for its manifestation and this is provided by the *dravya* (द्रव्य). So for better understanding of *guṇaḥ* (गुणः), we should have a knowledge of the *pāṁcabhautika* (पांचभौतिक) constitution of that particular *āśraya* (आश्रय) or *dravya* (द्रव्य).

We have already discussed earlier the fact that *Caraka* (चरक) in his *śārīrasthāna* (शारीरस्थान), has revealed that *laghutva* (लघुत्व) is due to the main presence of *ākāśa mahābhuta* (आकाश महाभुत) in its *pāṁcabhautika* (पांचभौतिक) constitution and furthermore it also incorporates significant attributes from *vāyu mahābhuta* (वायु महाभुत) and *agnī mahābhuta* (अग्नी महाभुत) apart from *laghu* (लघु) attribute.

As we are familiar with the effects of *mungo beans* (माष) and we know that it is in possession of *pṛthvi mahābhuta* (पृथ्वि महाभुत) and *jala mahābhuta* (जल महाभुत) in its *pāṁcabhautika* (पांचभौतिक) constitution, because it is used widely for the purpose of bṛṁhaṇa (बृंहण) activity.

So why the above graph is displaying the highest value of glycine in the *mungo beans* (माष). We will say that this graph is showing only half of the information about the *mungo beans* (माष). For full information about it, we should have a look at the following graph, which displays the comparative values of glutamic acid along with the glycine values of the selected food articles belonging to the *śimbīdhānya* (शिम्बीधान्य) group. The main thing it reveals that the *mungo beans* (माष) is very rich in amino acid values and it is not only in possession of higher values of *laghutīkṣṇoṣṇādi*

(लघुतीक्ष्णोष्णादि) as shown in earlier graph, but it also in better possession of *gururmandādi* (गुरुर्मन्दादि) attributes which are labeled as *jalamitra* (जलमित्र) groups of the *viṁśatī guṇāḥ* (विंशती गुणाः).

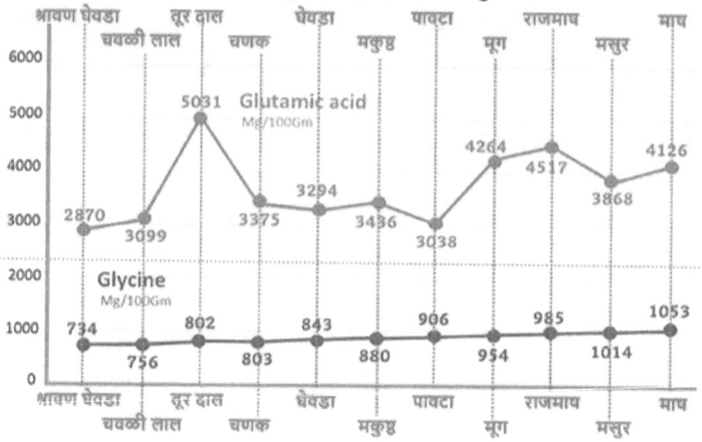

Here we can see that the glycine value of *mung beans* (माष) is 1053mg/100gm, and its glutamic acid value is 4126 mg/100gm, means the ratio of the glycine and glutamic acid values is nearly 1:4 in the *pāṁcabhautika* (पांचभौतिक) constitution of *mungo beans* (माष). French beans (श्रावण घेवडा), Adzuki beans (चवळी लाल), Bengal gram (चणक), Black beans (घेवडा), makuṣṭha (मकुष्ठ), Mung beans (मूग), Red lentil (मसुर) are also in possession of glycine and glutamic acid in the ratio of 1:4 in their *pāṁcabhautika* (पांचभौतिक) constitution. The ratio of the glycine and glutamic acid values of rājamāṣa (राजमाष) is 1:5, being slightly different from the above examples and makes it more prone to be in possession of *gururmandādi* (गुरुर्मन्दादि) attributes. Pigeon peas (तूर दाल) has also the same story, as its glycine value is 802 mg/100gm, and its glutamic acid value is 5031mg/100gm, maintaining the ratio of nearly 1:6 in the *pāṁcabhautika* (पांचभौतिक) constitution of Pigeon peas (तूर दाल).

It clearly explains that though the above graph is displaying the highest value of glycine in the *mungo beans* (माष), it should be examined in relation to its glutamic acid values.

In earlier pages, we have already discussed the two most important food ingredients of Indian diet, first one, godhūmaḥ (गोधूमः) and second one was Sorghum(देवधान्य), or generally known as jvārī (ज्वारी), and while understanding the *laghutīkṣṇoṣṇādi* (लघुतीक्ष्णोष्णादि) attributes of the Sorghum(देवधान्य) and different types of wheat (गोधूम), we have found that the ratio of glycine and glutamic acid in Sorghum(देवधान्य), appeared to be of 1:7 (glycine 346 mg/100gm: glutamic acid 2439 mg/100gm). So we had concluded that Sorghum (देवधान्य) is definitely in possession of *laghu* (लघु) and *sūkṣma guṇa* (सूक्ष्म गुण), while different types of wheat are showing the substantial excessive existence of *gururmandādi* (गुरुर्मन्दादि) attributes in comparison to the values of Sorghum(देवधान्य). Now while studying the *śimbīdhānya* (शिम्बीधान्य) group, we will consider the Sorghum (देवधान्य) as a base for comparison between the śūka varga (शूक वर्ग) and *śimbīdhānya varga* (शिम्बीधान्य वर्ग), for better understanding of *laghu* (लघु) attribute.

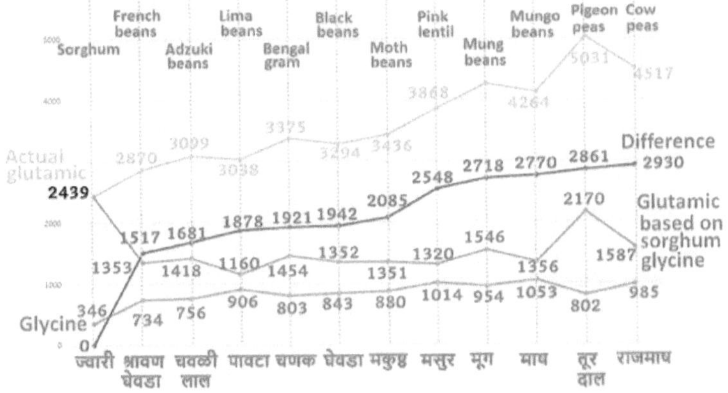

On the far left of the 'x' axis of the graph, we can see the "Sorghum or Jwari", which here we are considering as a base-reference for the glycine value. After it, all the names are belonging to the selected *śimbīdhānya varga* (शिम्बीधान्य वर्ग). The dark orange colored numbers and lines provide the information of the values of glycine belonging to the selected *śimbīdhānya* (शिम्बीधान्य), which range from 734mg/100gm to 1053mg/100gm. As discussed earlier, the values indicated by the green line and numbers are computer-generated by considering the ratio of glycine and glutamic acid values of the Sorghum (देवधान्य), and provide the information of the values of glutamic acid, belonging to the selected *śimbīdhānya* (शिम्बीधान्य), which are actually a projection of simulated values, depending upon the base value of glycine of the Sorghum (देवधान्य).

When we realize that the glycine value of French beans (श्रावण घेवडा) is 734 mg/100gms, at the same time, the actual glutamic acid of French beans (श्रावण घेवडा) is found to be 2870 mg/100gms, so we have to calculate, what will be the value of glutamic acid of French beans (श्रावण घेवडा), when associated with the glycine value of Sorghum (देवधान्य), and we can get it as 1353 mg /100 gms.

To find this answer, here we have generated a formula, which gives us the glutamic acid values of a substance when compared with the glycine value of Sorghum (देवधान्य).

$$\text{Glutamic acid value of śimbīdhānya (शिम्बीधान्य) compared with Glycine value of Sorghum} = \frac{\text{Glycine value of Sorghum} \times \text{Actual glutamic acid value of śimbīdhānya (शिम्बीधान्य)}}{\text{Glycine value of śimbīdhānya (शिम्बीधान्य)}}$$

	Glycine	Glutamic acid Actual	Glutamic acid (Based on Sorghum's glycine)	Difference
Sorghum	346	2439	2439	0
Shravan Ghevada	734	2870	1353	1517
Lal Chavali/Adzuki	756	3099	1418	1681
Vaal/Pavata	906	3038	1160	1878
Chana/Bengal gram	803	3375	1454	1921
Ghevada/Black beans	843	3294	1352	1942
Makushtha	880	3436	1351	2085
Maasha/Mungo beans	1053	4126	1356	2770
Rajmaasha/Chavali	985	4517	1587	2930

The third column in this table shows the values of glutamic acid, generated by the comparison with the glycine value of Sorghum (देवधान्य). The difference between the values of actual glutamic acid and the values of glutamic acid generated by calculation is displayed in the last column. As we have already established that the presence of higher values of glutamic acid is an indication of the availability of *gururmandādi* (गुरुर्मन्दादि) attributes in the entity. So when considering the values of Sorghum (देवधान्य) for comparison, we can find that if the difference between the values of actual glutamic acid and the values of glutamic acid generated by calculation is minimum, then there is a probability of scantiness or less finding the presence of *gururmandādi* (गुरुर्मन्दादि) qualities in that entity. In another word, that particular entity is prone to be less *guru* (गुरु) in its constitution, or we can also say that in that particular entity, *laghutīkṣṇoṣṇādi* (लघुतीक्ष्णोष्णादि) attributes should be present in abundant quantities. For example, here we can observe in the above table that the difference value of french beans or Shravan Ghevada is 1517mg/100gm, so the probability of the percentage of *laghu* (लघु) quality existent in the french beans should be high.

The average of the different values of the selected *śimbīdhānya* (शिम्बीधान्य), displayed in the last column of the table, is 2090mg/100gms. The difference between the values of actual glutamic acid and the values of glutamic acid generated by calculation in more than 75% of the substances from the selected *śimbīdhānya* (शिम्बीधान्य), noticeably shows that their values are below than the average value of the difference (<2259mg/100gms), revealing their propensity towards the likelihood of shortfall of *gururmandādi* (गुरुर्मन्दादि) qualities in these substances. Among these *śimbīdhānya* (शिम्बीधान्य), which are lesser in the difference values, makuṣṭha (मकुष्ठ) is in possession of difference value of 2085mg/100gms, indicating comparatively less significant percentage of *gururmandādi* (गुरुर्मन्दादि) qualities.

मकुष्ठो वातलो ग्राही कफपित्तहरो लघुः। भावप्रकाश धान्यवर्ग। ४९

In the dhānyavarga (धान्यवर्ग) described by bhāvaprakāśa (भावप्रकाश), makuṣṭha (मकुष्ठ) has been shown to be in possession of *laghu* (लघु) attribute, confirming our above discussion. Bengal gram or caṇaka (चणक) has a different value of 1921mg/100gms, which clearly reveals its tendency towards the possession of the *laghutīkṣṇoṣṇādi* (लघुतीक्ष्णोष्णादि) attributes, and it should be noted that bhāvaprakāśa (भावप्रकाश) has also described caṇaka (चणक) having in possession of *laghu* (लघु) attribute.

चणकः शीतलो रुक्षः पित्तरक्तकफापहः। लघुः कषायो विष्टम्भी वातलो ज्वरनाशनः। भावप्रकाश धान्यवर्ग। ४९

On the other hand, when the difference between the values of actual glutamic acid and the values of glutamic acid generated by calculation is higher, then there is higher possibility of finding more presence of *gururmandādi* (गुरुर्मन्दादि) qualities in that entity.

So relatively, here we can see that, as the difference value is increased, then the glutamic acid value should be found amplified and the *laghutīkṣṇoṣṇādi* (लघुतीक्ष्णोष्णादि) attributes should be present in limited extents in that entity. For example, the different values of the last two substances shown in the table, are in possession of higher values than the average difference value mentioned above. The highest difference value, in the given table of selected *śimbīdhānya* (शिम्बीधान्य), is 2930mg/100gms, which belongs to rājamāṣa (राजमाष), so we can say that rājamāṣa (राजमाष) is in possession of the highest value, which is responsible for the manifestation of *gururmandādi* (गुरुर्मन्दादि) attributes.

राजमाषो गुरुः स्वादुस्तुवरस्तर्पणः सरः। भावप्रकाश धान्यवर्ग। ४९

The second-highest difference value, in the given table of selected *śimbīdhānya* (शिम्बीधान्य), is 2770mg/100gms, which belongs to māṣa (माष) or Mungo beans, which reveals the fact that why māṣa (माष) is considered as in possession of *guru* (गुरु) attributes.

माषो गुरुः स्वादुपाकः स्निग्धो रुच्योऽनिलापहः।
भावप्रकाश धान्यवर्ग। ४९

	Glycine Actual	Glutamic Actual	Glycin based on glutamic acid of Sorghum	Difference
Sorghum	346	2439	346	0
Shravan Ghevada	734	2870	624	110
Lal Chavali	756	3099	595	161
Vaal- Pavata	906	3038	727	179
Black beans - ghevada	843	3294	624	219
Bengal gram	803	3375	580	223
Makushthka	880	3436	625	255
Masura	1014	3868	639	375
Mudga	954	4264	546	408
Tur Dal	802	5031	389	413
Maasha	1053	4126	622	431
Raj Maasha	985	4517	532	453

If we consider that the glutamic acid value of the Sorghum (देवधान्य) as a base for comparison, then the calculation formula will be as follows: -

$$\text{Glycine value of } \textit{śimbīdhānya} \text{ (शिम्बीधान्य) compared with Glutamic acid value of Sorghum} = \frac{\text{Glutamic acid value of Sorghum} \times \text{Actual glycine value of } \textit{śimbīdhānya} \text{ (शिम्बीधान्य)}}{\text{Glutamic acid value of } \textit{śimbīdhānya} \text{ (शिम्बीधान्य)}}$$

Where the actual value of glutamic acid of Sorghum (देवधान्य), is 2439 mg/100gms, then the actual value of glycine found in Sorghum (देवधान्य) is to be 346 mg/100gms, and the difference will be 0 mg/100gms. This is our base for the comparison.

One more important observation of this table is that when the difference value is increased, then the actual glutamic acid value of the concerned *śimbīdhānya* (शिम्बीधान्य) looks amplified. One should have a logic that why this significance should be taken into consideration. There are two reasons for this deliberation. First is enclosed in our earlier discussion, in which we have said that if there is an increase in glycine value of a substance, then there is a chance of an increased percentage for manifestation of *laghutīkṣṇoṣṇādi* (लघुतीक्ष्णोष्णादि) attributes in that entity, so one may think that as the difference value of glycine in the last column of the displayed table increases, there is a chance of being in possession of *laghu* (लघु) attribute, also upsurges. For example, as shown in the table, the last two substances in the table like māṣa (माष) or Mungo beans and rājamāṣa (राजमाष) are in possession of different values of glycine 431mg/100 gms and 453 mg/100 gms respectively, and their glycine values are also in high quantities. So one would think them as in possession of *laghu* (लघु) attribute. But the real situation is almost contrary to this thinking. The solution for this contradiction can be found in the second reason, where we have to observe the glutamic acid value along with the glycine. Here we can observe that as the difference in the glycine value is increased, the values of glutamic acid are also amplified. For example, while the difference in the values of actual glycine and calculated glycine of the french beans (श्रावण घेवडा) is 110mg/100 gms, the actual value of glutamic acid is 2870 mg/100gms. Like this, the māṣa (माष) and rājamāṣa (राजमाष) are in possession of highest difference values of glycine 431mg/100 gms and 453 mg/100 gms respectively, but their glutamic acid values are also found amplified as 4126mg/100 gms and 4517mg/100 gms correspondingly. This clearly indicates that only the difference in glycine values, should not be considered for the marking of *laghu* (लघु) attribute in that entity, and one should also think through the glutamic acid value, while this marking of *laghu* (लघु) attribute on any entity. As we can clearly see that from the selected *śimbīdhānya* (शिम्बीधान्य) in this table, no one is in possession of higher percentage of *laghu* (लघु) attribute than the quantity of *laghu* (लघु) attribute available in the Sorghum (देवधान्य), because their glutamic acid values are much higher than that of the Sorghum (देवधान्य). In simple way we can say that though the glycine values of māṣa (माष) and rājamāṣa (राजमाष) are highest in the given table, these two *śimbīdhānya* (शिम्बीधान्य) can not be said to be in possession of *laghu* (लघु) attribute in their constitution as their glutamic acid values are at more higher side, hence these two *śimbīdhānya* (शिम्बीधान्य) - māṣa (माष) and rājamāṣa (राजमाष) are considered to be in possession of *gururmandādi* (गुरुर्मन्दादि) attributes in comparison with the Sorghum (देवधान्य).

The average of the different values of the selected *śimbīdhānya* (शिम्बीधान्य), displayed in the last column of this table, is 293mg/100gms. The difference between the values of actual glycine and the values of glycine generated by calculation, in more than 50% of the substances from the selected *śimbīdhānya* (शिम्बीधान्य), strikingly indicates that their values are below than the average value of the difference (<293mg/100gms), but here we also have to consider the glutamic acid values of the concerned *śimbīdhānya* (शिम्बीधान्य). The values which are less than the average value of 293mg/100gms, are also in possession of lesser values of glutamic acid respectively. This consideration reveals their susceptibility towards the exhibition of presence of *laghutīkṣṇoṣṇādi* (लघुतीक्ष्णोष्णादि) attributes in that particular entity, in comparison with the other substances of the selected *śimbīdhānya* (शिम्बीधान्य). So somewhat, here we can perceive that, if the difference value of glycine is increased, and the glutamic acid value of the concerned *śimbīdhānya* (शिम्बीधान्य) is found to be in amplified quantity, then it is possible to be resulting in display of excessive *gururmandādi* (गुरुर्मन्दादि) qualities in that substance.

समाश्रयो अलनाईनः

Role of alanine in the manifestation of *laghu* (लघु) attribute: -

Until now, we have clearly understood the role of glycine in the foundation of manifestation of the *laghu* (लघु) attribute. Apart from this glycine amino acid, there are some more amino acids, which also take part in the foundation of the *laghu* (लघु) attribute. We will have a look at each of them one after another.

Another important one amino acid, after glycine is alanine. If we observe the structure of alanine, there are merely only carbon and hydrogen, therefore, alanine is considered as a hydrophobic amino acid.

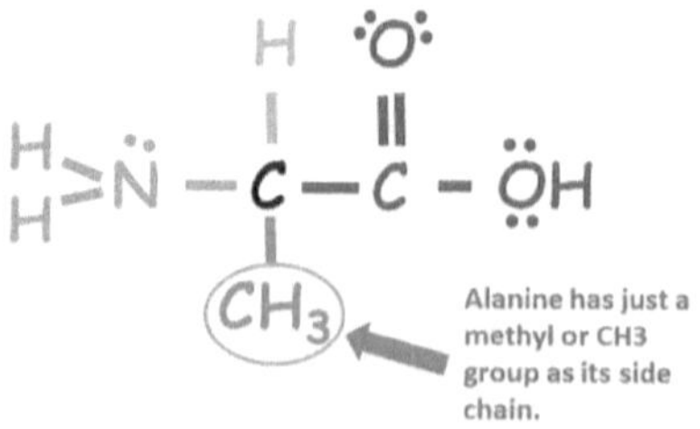
Alanine has just a methyl or CH3 group as its side chain.

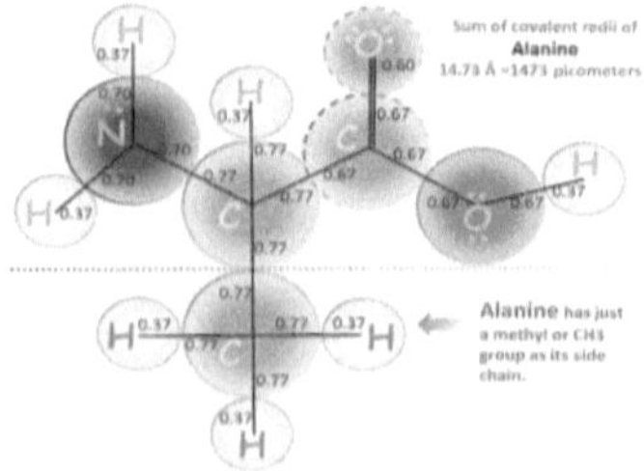
Sum of covalent radii of
Alanine
Alanine has just a methyl or CH3 group as its side chain.

The most important thing about alanine amino acid is that it is a tiny or miniature amino acid. This miniature size offers alanine, a capacity to enter in-between small spaces like a wedge. Dictionary meaning of the word 'wedge' is to force or press something into a space. This quality of alanine is produced or manifested due to the presence of *sukṣma* (सुक्ष्म) and *laghu* (लघु) attributes. On should note that, *vāgbhaṭṭa* (वाग्भट्ट) has described ākāśiya *dravya* (आकाशिय द्रव्य) as in possession of *sukṣma* (सुक्ष्म), and *laghu* (लघु) qualities and *carak* (चरक) has indicated *sūkṣma* (सूक्ष्म) along with *laghu* (लघु) qualities in the description of vāyaviya *dravya* (वायविय द्रव्य), and *vāgbhaṭṭa* (वाग्भट्ट) has described āgneya *dravya* (आग्नेय द्रव्य) as in possession of sūkṣma (सूक्ष्म) quality, so we can assume that alanine is in possession of three mahābhuta (महाभुत) - ākāśa (आकाश), vāyu (वायु), and agnī (अग्री), in its *pāṁcabhautika* (पांचभौतिक) constitution.

नाभसं सूक्ष्मविशदलघुशब्दगुणोल्बणम्। अष्टांगहृदय सूत्रस्थान ९-९

लघुशीतरुक्षखरविशदसूक्ष्मस्पर्श गुणबहुलानि वायव्यानि।

चरक सूत्र २६-११

रुक्षतीक्ष्णोष्णविशदसूक्ष्मरुपगुणोल्ब्णम्। आग्नेयं ... ॥

अष्टांग हृदय सूत्रस्थान ९-७

Earlier we have already discussed about the sūkṣma (सूक्ष्म) quality in details. As from these three *mahābhūta* (महाभूत), *any entity* acquires the ability to enter into smaller than the yocto particles, which is described as *sūkṣma guṇaḥ* (सूक्ष्म गुणः), hence, ḍalhaṇa (डल्हण) has described it as *sūkṣmamārgānupraveśī* (सूक्ष्ममार्गानुप्रवेशी) or having such penetrating power, which makes it capable or competent, to make a way through.

सूक्ष्ममार्गानुप्रवेशी। डल्हण २। १९-२०

विवरणे शक्तिः। हेमाद्रि १। १८

While discussing the function of *sūkṣma guṇa* (सूक्ष्म गुण), hemādri (हेमाद्रि) has also cited this power as "*vivaraṇe sūkṣmaḥ* (विवरणे सूक्ष्मः)". Thus we can say that along with *laghu guṇaḥ* (लघु गुणः), alanine also is in possession of *sūkṣma guṇaḥ* (सूक्ष्म गुणः), which is the reason that alanine is able to make its way through forcefully into the space of the molecules, which bears a resemblance to the above description given by hemādri (हेमाद्रि).

Before going further with our discussion, if we divert ourselves for a few minutes, from the main topic to alanine scanning, it will support us to enrich our understanding of alanine amino acid.

To define the role or involvement of a particular residue to the stability or function of a protein, alanine scanning is widely used in molecular biology. One of the benefits of the alanine scanning is that, with the help of mutagenesis, it can conclude the role of the side-chain of a specific residue in bioactivity. Mutagenesis means changing the genetic information of an organism, which becomes responsible for a mutation.

In the alanine scanning, residues in the target protein are substituted. Amongst the two reasons, that alanine becomes the substitution residue of choice, the first is that without changing the main-chain conformation and conserving the native protein structure, it takes away the side chains beyond the beta carbon and secondly, it does not allow to introduce steric or electrostatic effects.

Alanine scanning is a subtractive technique. Subtraction means taking away something from a group. Here alanine is used for replacing the target atom. In1989, the term alanine scanning was invented by Jim Wells in the mapping of the human growth hormone interaction surface with the corresponding receptor. High-resolution epitope mapping of hGH-receptor interactions by alanine-scanning mutagenesis. *Cunningham BC, Wells JA Science. 1989 Jun 2; 244(4908):1081-5.*

Surprisingly, the same technique used in alanine scanning was performed in the early seventies in India. As I remember, Vd. Antarkar, Ex-Director of Ayurveda, Maharashtra state, had conducted clinical and laboratory experiments for the study of ārogyavardhinī (आरोग्यवर्धिनी).

रस	गंधक	लोह	अभ्र	ताम्र	त्रिफला	शिलाजीत	गुग्गुल	चित्रक	
रस	गंधक	लोह	अभ्र	ताम्र	त्रिफला	शिलाजीत	गुग्गुल		कटुका
रस	गंधक	लोह	अभ्र	ताम्र	त्रिफला	शिलाजीत		चित्रक	कटुका
रस	गंधक	लोह	अभ्र	ताम्र	त्रिफला		गुग्गुल	चित्रक	कटुका
रस	गंधक	लोह	अभ्र	ताम्र		शिलाजीत	गुग्गुल	चित्रक	कटुका
रस	गंधक	लोह	अभ्र		त्रिफला	शिलाजीत	गुग्गुल	चित्रक	कटुका
रस	गंधक	लोह		ताम्र	त्रिफला	शिलाजीत	गुग्गुल	चित्रक	कटुका
रस	गंधक		अभ्र	ताम्र	त्रिफला	शिलाजीत	गुग्गुल	चित्रक	कटुका
रस		लोह	अभ्र	ताम्र	त्रिफला	शिलाजीत	गुग्गुल	चित्रक	कटुका
	गंधक	लोह	अभ्र	ताम्र	त्रिफला	शिलाजीत	गुग्गुल	चित्रक	कटुका

As shown in the above figure, in this experiment, he deducted every time one component from the herbo-mineral complex of ārogyavardhinī (आरोग्यवर्धिनी) and studied the effects of the remaining combination. Unfortunately, the documented evidence of these trials is not available today.

A question may arise that why alanine is used in the mutagenesis. The answer to this question is very important in the context of studying the *laghutva* (लघुत्व) of alanine. We are aware of the fact that secondary structures of proteins are polypeptide chains, which are created in such a way that make different shapes to perform a specific job, and are actually of two types, one is alpha-helix and second is beta-pleated sheet.

The alpha helix is designed, when the polypeptide chains twists into a spiral, facilitating all amino acids in the chain to create hydrogen bonds with each other. This hydrogen bond ties an oxygen molecule to a hydrogen molecule, permitting the helix to hold the spiral shape, which makes the alpha helix very strong. Alanine has a tendency to create alpha-helices and is normally equivalent to basically shortening by cutting off a side chain back to the beta carbon, which is the first side chain atom. Therefore, alanine is largely recognized single residue first choice for mutational scanning, because it retains the beta carbon but no other side chain functionality.

One should have a query, as we have already established that glycine is more in possession of *laghu* (लघु) and *sūkṣma guṇa* (सूक्ष्म गुण) in comparison with alanine, then why glycine is not used in alanine scanning in place of alanine. The answer itself denotes the difference between these two amino acids, as glycine, which takes away the beta carbon, is remarkably flexible and it can accept polypeptide backbone confirmations, usually not permissible by other amino acids, and this mutation results in flexibility and possible conformational changes. Subsequently, glycine will make the interpretations more complex in the experiment, so alanine is preferable to glycine in this scanning technique.

Now we will turn again towards our main discussion. There are some criteria, which we have already discussed at the time of glycine study, which are essential to be present in an entity before we can label the entity as in possession of *laghu guṇaḥ* (लघु गुणः). Now we will examine the alanine amino acid on these criteria.

First of all, we have established that alanine is in possession of *āakāśa mahābhūta* (आकाश महाभूत), *vāyu mahābhuta* (वायु महाभुत) and/or *agnī mahābhuta* (अग्नी महाभुत) in its *pāṁcabhautika* (पांचभौतिक) constitution.

As mentioned in *caraka* (चरक) **अवयवाः घटकात्मकद्रव्यम्। च वि १।१०**, we have also seen that the structure of alanine is very tiny and its side chain consists only of a CH_3.

As pointed out in "**परमाणुरुपाऽवयवांशाः। चरक शारीर ७।१७**", we have realized that, in the scales of atomic weight and atomic numbers of all hydrophobic amino acids, alanine's atomic number is 118 and its molecular mass is 71.08, and so it is placed at the second-lowest markings, making the structure of the alanine very tiny.

As described in " **अल्प अवयवत्वम् लघुत्वम् भार रहित्वम् वा।आयुर्वेददर्शन अध्याय ३-२-६-७** ", alanine has a methyl side-chain, which is non-reactive and rarely involved openly in any protein function. We have studied that the atomic radius of an element is the mean distance from the center of the nucleus to the boundary of the surrounding shells of electrons, which is majored in angstrom unit. Higher the values of distance, more bulky the atom and vice versa. Alanine has a covalent radius of 14.8 angstrom, which is second lowest value in the amino acids, which belongs to the "*jalaśatru* (जलशत्रु) " group of the *viṁśatī guṇāḥ* (विंशती गुणाः) described by *vāgbhaṭṭa* (वाग्भट्ट).

परमाणुरुपाऽवयवांशाः। चरक शारीर ७।१७

As discussed earlier, alanine is in possession of *agnī mahābhuta* (अग्नी महाभुत) in its *pāṁcabhautika* (पांचभौतिक) constitution, resulting in being capable of increasing the jāṭharāgnī (जाठराग्नी)/ dhātvāgnī (धात्वाग्नी).

सूक्ष्मः आग्नेयद्रव्यस्य गुणेष्वेकः। सुश्रुत सूत्र ४१।३

As alanine is able to make its way through into the space of the molecules forcefully, due to the presence of *sukṣma* (सुक्ष्म) and *laghu* (लघु) attributes, it helps the physiological transformation, breakdown process of the entity very easy and simple.

शीघ्रपाकी। चरक शा। ६।१०

अलनाईन तथा मांस वर्ग

In sequence, before labeling the alanine amino acid to be in possession of a *laghu* (लघु) and *sūkṣma* (सूक्ष्म) attribute, one more significant condition, is needed to be checked. We have held earlier that main activity of such entities, should be application of *laghutva* (लघुत्व) in the body, as described "**अस्य लङ्घने शक्तीः। अष्टांग हृदय सूत्र १।१८** ", we should capable to demonstrate the ability of alanine to be valuable as a member of food constituents in the weight loss activity.

Like the method, which we used in the understanding of glycine, here also we will discover the *māṁsa varga* (मांस वर्ग) and *śimbīdhānya varga* (शिम्बीधान्य वर्ग), for the better understanding of the *laghu* (लघु) and *sūkṣma* (सूक्ष्म) attribute of alanine.

For distinguishing the *laghutva* (लघुत्व) of the *māṁsa dhātu* (मांस धातु) of the wild-life belonging to the *mṛga* (मृग) group, in relation with the *māṁsa dhātu* (मांस धातु) of the faunae of *mahāmṛga* (महामृग) group, we must match the amino acids of the meat belonging to both groups.

In the subsequent discussion, we will perceive different amino acids related to the *jalaśatru* (जलशत्रु) group, deliberated earlier.

For the argument, we have selected nine types of *māṁsa* (मांस) of the wild-life, which belong to the group of *mṛga* (मृग) such as, Deer raw (अपक्व मृगमांस), Deer shoulder roasted Venison (हरिणपक्ष भर्जितः), Deer cooked roasted (हरिण भृष्टः), Deer Ground Raw Venison (अपक्व सारंग), Rabbit wild raw (अपक्व वन शशः), Rabbit domestic raw (अपक्व ग्राम्य शशः), Rabbit wild cooked (वन

शशः पाचितः), Rabbit domestic stewed (ग्राम्य शशः क्वाथितः), and Rabbit domestic roasted (ग्राम्य शशः भर्जितः).

For the purpose of comparison, we have designated nine types of *māṁsa* (मांस) of the wild-life, which belong to the *mahāmṛga* (महामृग) group, such as, Beef raw (वनगव अपाचितः), Bison *Gava* (वन्य धेनु), Bison- *Reda* (वन्य गौर), Boar wild cooked (वन्य वराह पाचितः भृष्टः), Beef brisket (गोवत्स पाचितः), Buffalo roasted (महिष पाचितः भृष्टः), Veal raw (गोवत्स अपाचितः), Boar wild raw (वन्य वराह अपाचितः), Pork domestic raw (ग्राम्य वराह अपाचितः), Buffalo raw (महिष अपाचितः), Pork cured ham patties (वराह पाचितः), Veal- trimmed cuts cooked (गोवत्स त्वष्ट्र पाचितः भृष्टः), Veal- trimmed cuts raw (गोवत्स त्वष्ट्र अपाचितः).

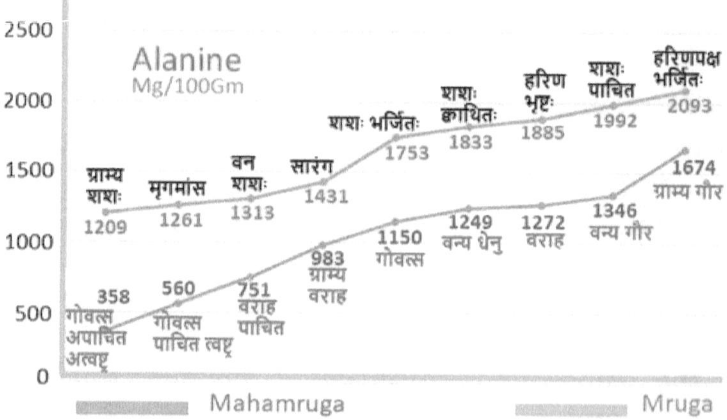

Table showing the high levels of Alanine in *mṛga* (मृग) group.

While observing the overhead table, we can discover that, the brown line presenting the alanine quantity of wild-life meat, belonging to the *mṛga* (मृग) group, is unquestionably with greater values than the values of blue line, which are suggestive of the alanine qualities of wildlife meat belonging to *mahāmṛga* (महामृग) group. We have previously conversed that, the existence of alanine in a molecule validates the manifestation of *vāyu mahābhūta* (वायु महाभूत), *agni mahābhūta* (अग्नि महाभूत) and *āakāśa mahābhūta* (आकाश महाभूत), which are indispensable components of the *laghu* (लघु) attribute. In previous discussion, we have already seen the inspiration or effect of *jāṅagalav deśa* (जाङ्गल देश), *ānupa deśa* (आनुप देश) and *sādhāraṇa deśa* (साधारण देश) in the manifestation of attributes, particularly of *laghu* (लघु) quality. We have also seen that the *kāla* (काल) or season of the life/year, has also a particular role while defining the qualities of any entity. *vāgbhaṭṭa* (वाग्भट्ट) has classified this impact of *kāla* (काल) in three groups, and hemādri (हेमाद्रि) in his commentary has shown the fact that *ādirbālyāvasthā kaphasya* (आदिर्बाल्यावस्था कफस्य) means in the childhood *kapha* (कफ) is more noticeable than other *doṣa* (दोष).

वयोहोरात्रिभुक्तानां तेऽन्तमध्यादिगाः क्रमात्। अष्टांग हृदय सूत्र।१।८ दोषकालानाह-वय इति। तत्र वय शरीर परिणाम ....आदिर्बाल्यावस्था कफस्य। आयुर्वेद रसायन- हेमाद्रि।

As *kapha doṣa* (कफ दोष) is in custody of *pṛthvi mahābhuta* (पृथ्वि महाभुत) and *jala mahābhuta* (जल महाभुत) in their *pāṁcabhautika* (पांचभौतिक) constitution, evidently, all the living entities on the planet are having *pṛthvi mahābhuta* (पृथ्वि महाभुत) and *jala mahābhuta* (जलमहाभुत) principally in their *pāṁcabhautika* (पांचभौतिक) constitution at the time of their childhood, because as revealed by hemādri (हेमाद्रि), *bālyāvasthā* (बाल्यावस्था) is generally ruled by *kapha doṣa* (कफ दोष).

When we say that at *bālyāvasthā* (बाल्यावस्था), there is a prevalence of *pṛthvi mahābhuta* (पृथ्वि महाभुत) and *jala mahābhuta* (जल महाभुत) in the *pāṁcabhautika* (पांचभौतिक) constitution of the body, the hidden significance is that, other three mahābhuta (महाभुत) -

ākāśa (आकाश), vāyu (वायु), and agnī (अग्नी), have a secondary role to carry out in the *bālyāvasthā* (बाल्यावस्था), and they are ineffective to prove their presence in the manifestation of the attributes. This phenomenon leads to the reduced exhibition of *rukṣasūkṣmādi* (रुक्षसूक्ष्मादि) qualities in the body.

The determination of all this discussion leads us in the direction of the assumption that, in the *bālyāvasthā* (बाल्यावस्था), *gururmandsnigdhādi* (गुरुर्मन्दस्निग्धादि) qualities are consistently surplus in quantities, and *rukṣasūkṣmādi* (रुक्षसूक्ष्मादि) qualities are in much less extents in the body.

If again, we have a look at the graph displaying the levels of alanine presented above, we can notice that, the last two lowest numbers (358 mg/100gms and 560 mg/100gms) from the values of blue line, which are suggestive of the alanine quantities of wild-life meat belonging to *mahāmṛga* (महामृग) group, are designated to the flesh of Veal. Earlier we have seen that Veal means meat of calves and beef specifies meat of grown-up cattle. Generally after attainment to 18 to 20 weeks of the age, and when weight is nearly 200 kg, calves are slaughtered. Even though the natural life span is of 15-20 years, usual slaughter age of "Veal" calves is 1-24 weeks, which clearly shows that, calves are slaughtered when they are in *bālyāvasthā* (बाल्यावस्था), and at this period of life, *gururmandsnigdhādi* (गुरुर्मन्दस्निग्धादि) qualities in their body, are always prevailing, in comparison to the *rukṣasūkṣmādi* (रुक्षसूक्ष्मादि) qualities in their body. Therefore it is not surprising that the values of alanine look relatively less in the above table. Our study has shown that in Veal flesh, the *gururmandsnigdhādi* (गुरुर्मन्दस्निग्धादि) qualities are approximately nearly three times more than the *rukṣasūkṣmādi* (रुक्षसूक्ष्मादि) qualities present in it. One more reflection about these two values of Veal flesh is that 560mg/100gm is the value of cooked and trimmed Veal cuts, whereas 358 mg/100gm is the value of raw Veal flesh. This variance in these two values is due to the effect of cooking the Veal, of course, the result of usage of agnī (अग्नी).

Considering the effect of agnī (अग्नी) on the values of raw flesh, we can observe that the values of alanine increases after the application of agnī (अग्नी), as shown in the following table: -

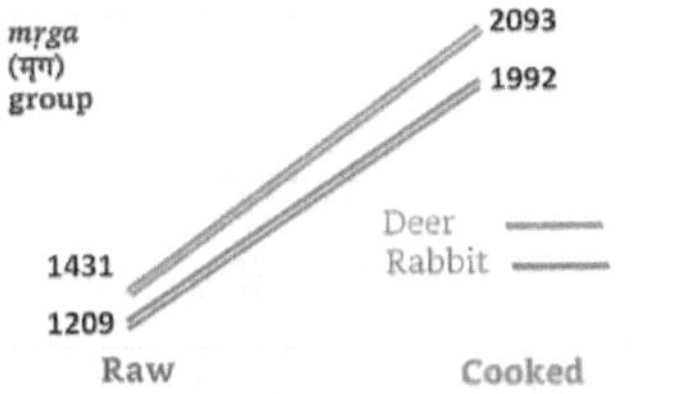

Table showing the change in Alanine values after processing

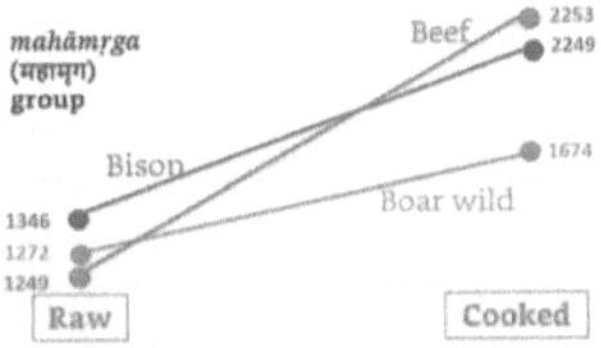

Table showing the change in **Alanine** values after processing

On the left side of the table, we can observe the values of alanine in the flesh of two animals from *mṛga* (मृग) group, in which the alanine value of the flesh of rabbit has augmented by 64.76%, while the alanine value of the flesh of deer has improved by 46.26%.

On the right side of the table, these are the meats of three animals from *mahāmṛga* (महामृग) group, in which the alanine values of the flesh of wild boar, bison, and beef have been shown to be increased by 31.60%, 67.08% and 80.38% correspondingly.

These differences are due to the application of agni in the varied processings for the cooking, which we will discuss later in the description of sūkṣma guṇa (सूक्ष्म गुण) and agnī (अग्नी).

We have seen that there are 3 sub-groups of *jāṅagala* (जाङ्गल) type of wild-life animals such as *mṛga* (मृग), *viṣkira* (विष्किर), and *pratuda* (प्रतुद).

आद्यास्त्रयो मृगविष्किरप्रतुदा - जाङ्गला।

In earlier pages, we have conversed in particulars about the flesh of *mṛga* (मृग) type of wildlife. Now we will have a look at the *viṣkira* (विष्किर), and *pratuda* (प्रतुद) types of wildlife.

लाववार्तीकवर्तीररक्तवर्त्मककुक्कुभाः।

कपिञ्जलोपचक्राख्यचकोरकुरुबाहवः॥४४

वर्तको वर्तिका चैव तित्तिरिः क्रकरः शिखि।

ताम्रचूडाख्यबकरगोनर्दगिरिवर्तिकाः॥४५

तथा शारपदेन्द्राभवरटाद्याच विष्किरा। अष्टांग हृदय सूत्र ६

For the purpose of clear understanding, we have carefully chosen five types of flesh of the animals, which belong to the group of *viṣkira* (विष्किर) such as, lāvā (लावा), kapota (कपोत), haṁsa (हंस), tāmracūḍa or kukkuṭa (ताम्रचूड तथा कुक्कुट), and bhāradvāja (भारद्वाज).

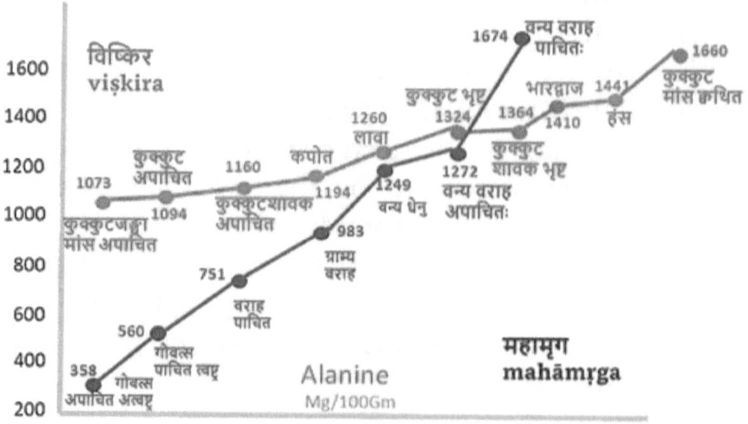

This graph illustrates the variance in the values of alanine among the flesh belonging to the *mahāmṛga* (महामृग) and *viṣkira* (विष्किर) wild-life animals.

In the group of *viṣkira* (विष्किर), the brown-colored names and figures are indicative of these animals which are rarely consumed nowadays. In some part of the nation, like a north-east area, till birds belonging to the *viṣkira* (विष्किर) group of wild-life are utilized as a part of daily food, but majority these brown colored names are out of the spectra of our food. But the figures and names given in pink color are a regular part of Indian food. We have specially mentioned six types of meat of chicken which are consumed in a vast proportion of mass.

kukkuṭaśāvaka (कुक्कुटशावक) or a young chicken, after processed by either roasting, grilling or barbecuing, becomes more *laghu* (लघु) in its nature due to the increase in the qualities belonging to "*jalaśatru* (जलशत्रु) " group of the *viṁśatī guṇāḥ* (विंशती गुणाः) described by *vāgbhaṭṭa* (वाग्भट्ट). We can observe in the above graph that the raw kukkuṭaśāvaka (कुक्कुटशावक) is in possession of the alanine value of 1160 mg/100 gm, while after cooking, the alanine value is enhanced up to 1364 mg/100 gm. This upsurge of 204 mg/100 gm numbers in the value of alanine, reveals the intensification of *laghu* (लघु) quality in the processed kukkuṭaśāvaka (कुक्कुटशावक). In the matter of matured kukkuṭa (कुक्कुट), the raw bird gives the value of alanine as 1073 mg/100 gm, while after cooked slowly in liquid, in a closed dish or pan, the meat of the kukkuṭa (कुक्कुट) turns out to be in possession of 1660 mg/100 gm of alanine, with nearly 54.70% rise in the alanine quantity. This reveals the fact that though the birds belonging to the Phasianidae family, are basically in possession of more *laghu* (लघु) quality, in comparison with the animals of *mahāmṛga* (महामृग) group of animals, if these chicken birds are cooked slowly on low heat for a prolonged period, they become transformed having more *laghu* (लघु) quality in their possession.

Here we can perceive that, the pink and brown mixed colored line, showing the alanine quantities of wild-life meat, belonging to the *viṣkira* (विष्किर) group, is unquestionably with greater values than the values of blue line, which is suggestive of the alanine qualities of wildlife meat having its place in *mahāmṛga* (महामृग) group. The average value of the alanine from the flesh of wild-life of *viṣkira* (विष्किर) group is 1298 mg/100 gm, while the average value of alanine of the flesh of animals belonging to *mahāmṛga* (महामृग) group is found to be 978 mg/100 gm only. Therefore, it obviously exhibits that, in the given sample the meat of the animals from *viṣkira* (विष्किर) group are in possession of 32% more alanine, than the alanine available in the meat of animals from *mahāmṛga* (महामृग) group. Now it has been previously proved in earlier discussion that the existence of alanine in a molecule, establishes the occurrence of *vāyu mahābhūta* (वायु महाभूत), *agni mahābhūta* (अग्नि महाभूत) and *āakāśa mahābhūta* (आकाश महाभूत), and these three *mahābhūta* (महाभूत) are actually fundamental ingredients of the *laghu* (लघु) attribute.

शिम्बीवर्गीय अलनाईन तथा लघुत्वम्

Understanding alanine in śimbīdhānya varga (शिम्बीधान्य वर्ग): -

At the time of study of alanine amino acid, we have decided that the easy way to comprehend the particulars of any *dravya* (द्रव्य), is to observe its *pāṁcabhautika* (पांचभौतिक) constitution.

Therefore, now we will match the structures of some of the food article belonging to *śimbīdhānya varga* (शिम्बीधान्य वर्ग). Actually, the popular and most common Indian vegetarian diet originates from two major groups - śūka varga (शूक वर्ग) and *śimbīdhānya varga* (शिम्बीधान्य वर्ग), which are defined in the *annasvarupa vijñānīya* (अन्नस्वरुप विज्ञानीय) chapter by *vāgbhaṭṭa* (वाग्भट्ट).

In earlier chapter, we have debated that wheat or godhūmaḥ (गोधूमः) as in possession of *guru* (गुरु), *snigdha* (स्निग्ध), and *sthira* (स्थिर) qualities, and these 3 attributes come under the classification of " *jalamitra* (जलमित्र)", apparently *laghutīkṣṇoṣṇādi* (लघुतीक्ष्णोष्णादि) attributes should be present with less magnitudes in wheat or godhūmaḥ (गोधूमः). We have already conversed that alanine is definitely in possession of *laghu* (लघु) and *sūkṣma guṇa* (सूक्ष्म गुण). Accordingly while reviewing the *laghutīkṣṇoṣṇādi* (लघुतीक्ष्णोष्णादि) attributes, we should always have an eye on the values of alanine also. Even we can say that minimum the value of alanine, greater the chances of the entity being in possession of *gururmandādi* (गुरुर्मन्दादि) attributes and higher the value of alanine, then there are chances of the entity being in possession of *laghutīkṣṇoṣṇādi* (लघुतीक्ष्णोष्णादि) attributes.

Apart from wheat or godhūmaḥ (गोधूमः), roṭī (रोटी) or bhākarī (भाकरी) prepared from the flour of jvārī (ज्वारी) or Sorghum(देवधान्य), is one of the principal sources of nutrition in rural India. We have earlier established that wheat or godhūmaḥ (गोधूमः) is in possession of more glutamic acid than that of jvārī (ज्वारी) or Sorghum(देवधान्य).

Understanding of the *laghutīkṣṇoṣṇādi* (लघुतीक्ष्णोष्णादि) attributes of the Sorghum(देवधान्य) and different types of wheat (गोधूम), in context with the alanine amino acid, after studying the ratio of alanine and glutamic acid in Sorghum(देवधान्य), it appears to be of 1:2.4 (alanine 1033 mg/100gm: glutamic acid 2439 mg/100gm). If we assume that there is zero difference in between the *laghutīkṣṇoṣṇādi* (लघुतीक्ष्णोष्णादि) attributes of Sorghum(देवधान्य) and different types of wheat (गोधूम), then the values of alanine and glutamic acid should be in the ration of 1:2.4 as shown in the following table.

Comparison of Alanine and Glutamic acid of Sorghum and different Wheats

	Sorghum	Wheat-1	Wheat-2	Wheat-3	Wheat-4
Alanine	1033	0427	0487	0450	0555
Glutamic acid as per ratio	2439	1008	1149	1062	1310
Actual Glutamic acid	2439	4743	4324	3998	4946
Difference	0000	3735	3175	2936	3636

Base of calculation is Alanine : Glutamic acid ratio of Sorghum = 1.00 : 2.36

We can see that there is a vast difference in the actual values as shown in the above table, which reveals that Sorghum(देवधान्य) is absolutely in possession of *laghu* (लघु) and *sūkṣma guṇa* (सूक्ष्म गुण), while different types of Wheats are showing the considerable extreme presence of *gururmandādi* (गुरुर्मन्दादि) attributes in comparison to the values of Sorghum(देवधान्य).

Comparing to the alanine and glutamic values of Sorghum (देवधान्य) and all types of godhūmaḥ (गोधूमः), we can determine that Sorghum(देवधान्य) has less nutritional values than all types of wheat due to the presence of alanine also.
Thus far, we have debated two significant entities - godhūmaḥ (गोधूमः) and Sorghum (देवधान्य), which is included under the classification of śūka varga (शूक वर्ग), and their percentage of consumption in Indian diet is too high to neglect.
Here and now we will move further, in the direction of our study of the next group of the food articles, which belong to *śimbīdhānya varga* (शिम्बीधान्य वर्ग).

मुद्गाढकीमसूरादि शिम्बिधान्यं विबन्धकृत्।

कषायं स्वादु सङ्ग्राहि कटुपाकं हिमं लघु॥अष्टांग हृदय सूत्र ६।१७

For the study of alanine we have selected green peas (कलाय), french beans (श्रावण घेवडा), adzuki beans (लाल चवळी), pigeon peas (तूर डाल), *caṇaka* (चणक), black beans (घेवडा), *makuṣṭha* (मकुष्ठ), lima beans (पावटा), mung beans (मूग), rajmaasha (राजमाष), *masura* (मसुर) and *māṣa* (माष) from the *śimbīdhānya varga* (शिम्बीधान्य वर्ग).

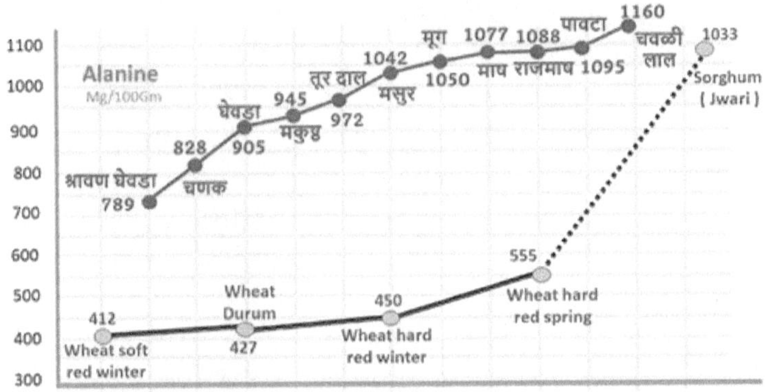

When we try to understand the blue line and figures in the above chart, we can see that the alanine value of Sorghum(देवधान्य), which is 1033 mg/100gm, and the alanine values of winter wheats (गोधूम रब्बी)- soft red, durum, hard red winter and hard red spring are 412, 427, 450, and 555 mg /100 gm correspondingly. A glimpse at the graph discloses that alanine values of all types of wheat are absolutely in lower quantity than the alanine values of the selected *śimbīdhānya* (शिम्बीधान्य), which are presented in the graph by purple line and figures. One interesting thing is that nearly 6 *śimbīdhānya* (शिम्बीधान्य) are in possession of higher alanine values than the alanine value of Sorghum(देवधान्य), which is 1033 mg/100gm. So one will think that *mungo beans* (माष) and rajmaasha (राजमाष) having alanine value of 1077 mg/100gm and 1088 mg/100gm respectively, should be in possession of *laghu* (लघु) attribute, because we have already discussed that when there is higher percentage of alanine present in the entity, then it should possess *laghu* (लघु) attribute. But this is not true, as we know that they are in possession of *pṛthvi mahābhuta* (पृथ्वि महाभुत) and *jala mahābhuta* (जल महाभुत) in their *pāṁcabhautika* (पांचभौतिक) constitution, and are used widely for the purpose of bṛṁhaṇa (बृंहण) activity. The reason, why the above graph is displaying higher values of alanine of the *mungo beans* (माष) and rajmaasha (राजमाष) is hidden in the following graph.

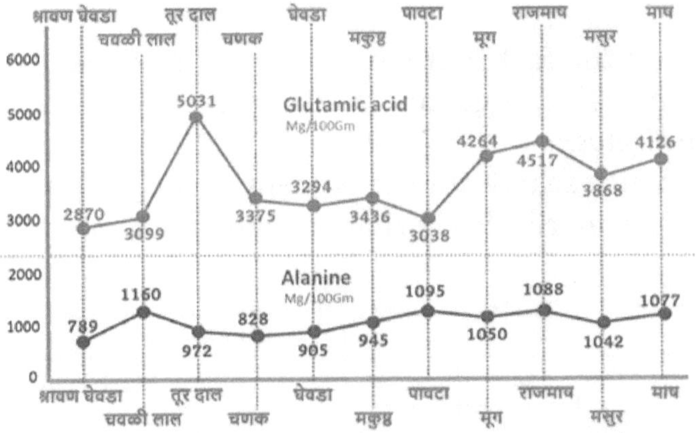

Here in this graph, the comparative values of glutamic acid along with the alanine values of the selected food articles belonging to the the *śimbīdhānya* (शिम्बीधान्य) group are displayed, which reveals that the *mungo beans* (माष) is very rich in amino acid values and it is not only in possession of higher values of *laghutīkṣṇoṣṇādi* (लघुतीक्ष्णोष्णादि) as shown in earlier graph, but it also in better possession of *gururmandādi* (गुरुर्मन्दादि) attributes.

Here we can see that the alanine value of *mung beans* (माष) is 1077mg/100gm, and its glutamic acid value is 4126 mg/100gm, means the ratio of the alanine and glutamic acid values is nearly 1.00: 3.83 in the *pāṁcabhautika* (पांचभौतिक) constitution of *mungo beans* (माष). Even bengal gram (चणक), Mung beans (मूग), rajmaasha (राजमाष), tūra dāla (तूर दाल) are in possession of alanine and glutamic acid in the ratio of nearly 1.00: 4.4 in their *pāṁcabhautika* (पांचभौतिक) constitution. Therefore, though there are higher values of alanine in these selected *śimbīdhānya* (शिम्बीधान्य), the ratio of the alanine and glutamic acid values of these entities clearly shows that the higher percentage of glutamic acid has made them more prone to be in possession of *gururmandādi* (गुरुर्मन्दादि) attributes. This also makes us aware of the fact that we should not be dependent on only the value of glycine or alanine for the decision of *laghutīkṣṇoṣṇādi* (लघुतीक्ष्णोष्णादि) attributes but we should also search for the percentage of *gururmandādi* (गुरुर्मन्दादि) attributes in that particular entity.

	Alanine Actual	Glutamic Actual	Glutamic acid based on Sorghum alanine	Difference
Shravan Ghevada	789	2870	3758	-888
Bengal gram	828	3375	4211	-836
Black beans - Ghevada	905	3294	3760	-466
Makushtha	945	3436	3756	-320
Tur Dal	972	5031	5347	-316
Sorghum	1033	2439	2439	0
Masura	1042	3868	3835	33
Mudga	1050	4264	4195	69
Maasha	1077	4126	3957	169
Raj Maasha	1088	4517	4289	228
Vaal- Pavata	1095	3038	2866	172
Lal Chavali	1160	3099	2760	339

$$\text{Glutamic acid value of } \textit{śimbīdhānya} \text{ (शिम्बीधान्य) compared with Alanine value of Sorghum} = \frac{\text{Alanine value of Sorghum} \times \text{Actual glutamic acid value of } \textit{śimbīdhānya} \text{ (शिम्बीधान्य)}}{\text{Alanine value of } \textit{śimbīdhānya} \text{ (शिम्बीधान्य)}}$$

In this table, we are considering the alanine value of Sorghum (देवधान्य) as a base for comparison between the śūka varga (शूक वर्ग) and *śimbīdhānya varga* (शिम्बीधान्य वर्ग), for better understanding of *laghu* (लघु) attribute, so its difference value is zero.

The second column of the above table provides the figures of the values of actual alanine, belonging to the selected *śimbīdhānya* (शिम्बीधान्य), which range from 789mg/100gm to 1160mg/100gm. The values exhibited in the fourth column, are computer-generated by bearing in mind the ratio of alanine and glutamic acid values of the Sorghum (देवधान्य), and makes available the information of the values of glutamic acid, belonging to the selected *śimbīdhānya* (शिम्बीधान्य), which are essentially a projection of simulated values, depending upon the base value of alanine of the Sorghum (देवधान्य). Once we recognize that the alanine value of French beans (श्रावण घेवडा) is 789 mg/100gms, at the same time, the actual glutamic acid of French beans (श्रावण घेवडा) is found to be 2870 mg/100gms, so we have to compute, what will be the value of glutamic acid of French beans (श्रावण घेवडा), when related with the alanine value of Sorghum (देवधान्य), and we can get it as 3758 mg /100 gms, with the help of formula given at the end of above table.

The average of the difference of the selected *śimbīdhānya* (शिम्बीधान्य) values, showed in the last column of the table, is -165 mg/100gms. The variance between the values of actual glutamic acid and the values of glutamic acid generated by calculation in more than 46% of the entities from the selected *śimbīdhānya* (शिम्बीधान्य), prominently indicates that their values are below than the average value of the difference (<-165mg/100gms), illuminating their tendency towards the likelihood of shortfall of *gururmandādi* (गुरुर्मन्दादि) qualities in these substances. On the other hand more than 54% of the entities from the selected *śimbīdhānya* (शिम्बीधान्य), prominently indicates that their values are more than the average value of the difference (<-165 mg/100gms), revealing the fact that these entities are in possession of *gururmandādi* (गुरुर्मन्दादि) qualities.

Between these *śimbīdhānya* (शिम्बीधान्य), which are lesser in the difference values, makuṣṭha (मकुष्ठ) is in possession of difference value of -320mg/100gms, representative of less significant percentage of *gururmandādi* (गुरुर्मन्दादि) qualities. As seen earlier, in the dhānyavarga (धान्यवर्ग) described by bhāvaprakāśa (भावप्रकाश), makuṣṭha (मकुष्ठ) has been shown to be in possession of *laghu* (लघु) attribute.

मकुष्ठो वातलो ग्राही कफपित्तहरो लघुः। भावप्रकाश धान्यवर्ग। ४९

Bengal gram or caṇaka (चणक) has a difference-value of -836 mg/100gms, which undoubtedly discloses its tendency towards the possession of the *laghutīkṣṇoṣṇādi* (लघुतीक्ष्णोष्णादि) attributes, and it should be noted that bhāvaprakāśa (भावप्रकाश) has also described caṇaka (चणक) having in possession of *laghu* (लघु) attribute.

चणकः शीतलो रुक्षः पित्तरक्तकफापहः।

लघुः कषायो विष्टम्भी वातलो ज्वरनाशनः। भावप्रकाश धान्यवर्ग। ४९

On the other hand, when the difference between the values of actual glutamic acid and the values of glutamic acid generated by calculation is higher, then there is higher possibility of finding more presence of *gururmandādi* (गुरुर्मन्दादि) qualities in that entity. For example, the difference-values of the last four substances presented in the table, are in possession of greater values than the average difference value mentioned above. The difference values of māṣa (माष) and rājamāṣa (राजमाष) in the given table of selected *śimbīdhānya* (शिम्बीधान्य), are 228mg/gms and 339mg/100gms respectively. Therefore, we can say that māṣa (माष) and rājamāṣa (राजमाष) are in possession of the alanine value in such quantity, which becomes responsible for the manifestation of *gururmandādi* (गुरुर्मन्दादि) attributes.

माषो गुरुः स्वादुपाकः स्निग्धो रुच्योऽनिलापहः।
भावप्रकाश धान्यवर्ग। ४९

राजमाषो गुरुः स्वादुस्तुवरस्तर्पणः सरः। भावप्रकाश धान्यवर्ग। ४९

This can be calculated by one more way to understand the *laghu* (लघु) or *guru* (गुरु) attribute of selected *śimbīdhānya* (शिम्बीधान्य) by changing the base for the comparison. In spite of alanine value of the Sorghum (देवधान्य), if we consider the glutamic acid value of the Sorghum (देवधान्य) as a base for comparison, then the calculation formula will be as follows: -

$$\begin{array}{l}\text{Alanine value of}\\ \textit{śimbīdhānya}\ \text{(शिम्बीधान्य)}\\ \text{compared with}\\ \text{Glutamic acid value of}\\ \text{Sorghum}\end{array} = \frac{\text{Glutamic acid value of Sorghum} \times \text{Actual Alanine value of } \textit{śimbīdhānya}\ \text{(शिम्बीधान्य)}}{\text{Glutamic acid value of } \textit{śimbīdhānya}\ \text{(शिम्बीधान्य)}}$$

Where the actual value of glutamic acid of Sorghum (देवधान्य), is 2439 mg/100gms, then the actual value of alanine found in Sorghum (देवधान्य) is to be 1033 mg/100gms, and the difference will be 0 mg/100gms. This is our base for the comparison.

	Alanine Actual	Glutamic Actual	Alanine based on Sorghum Glutamic acid	Difference
Sorghum	1033	2439	1033	000
Shravan Ghevada	789	2870	671	118
Vaal- Pavata	1095	3038	879	216
Bengal gram	828	3375	598	230
Black beans - Ghevada	905	3294	670	235
Lal Chavali	1160	3099	913	247
Makushtha	945	3436	671	274
Masura	1042	3868	657	385
Maasha	1077	4126	637	440
Mudga	1050	4264	601	449
Raj Maasha	1088	4517	587	501
Tur Dal	972	5031	471	501

While observing this table carefully, we can fond that when the difference value is in increasing state, then the actual glutamic acid value of the concerned *śimbīdhānya* (शिम्बीधान्य) appears to be amplified. The significance of this phenomenon should be considered properly. The main deliberation behind this is discussed earlier, in which we have supposed that if there is an enhancement in alanine value of a substance, then there is a chance of an increased percentage for manifestation of *laghutīkṣṇoṣṇādi* (लघुतीक्ष्णोष्णादि) attributes in that entity, so one will think it as the difference value of alanine in the last column of the displayed table increases, there should be a chance for an entity of being in possession of *laghu* (लघु) attribute. For example, as shown in the table, the last two substances in the table like rājamāṣa (राजमाष) and tūra dāla (तूर दाल) are in possession of difference-values of alanine 501mg/100 gms both, and their alanine values are also in high quantities (1088mg/100 gms and 972 mg/100 gm). So one would consider, both these are in possession of *laghu* (लघु) attribute. But the factual state is almost conflicting to this philosophy. The explanation for this ambiguity can be found in the second phenomenon, where we have to observe the glutamic acid value along with the alanine. Here we can see that as the difference in the alanine value is increased, the values of glutamic acid are also amplified. For example, while the difference in the values of actual alanine and calculated alanine of the french beans (श्रावण घेवडा) is 118mg/100 gms, the real value of glutamic acid is 2870 mg/100gms. Like this, the māṣa (माष) and rājamāṣa (राजमाष) are in possession of highest difference values of alanine 440mg/100 gms and 501 mg/100 gms respectively, but their glutamic acid values are also found amplified as 4126mg/100 gms and 4517mg/100 gms respectively.

This obviously points out that only the difference in alanine values, should not be substantial for the marking of *laghu* (लघु) attribute in that entity, and one should also think through the glutamic acid value, while this labeling of *laghu* (लघु) attribute on any entity.

In earlier discussion in the study of glycine we had said that from the selected *śimbīdhānya* (शिम्बीधान्य) in the table related with glycine, no one is in possession of higher percentage of *laghu* (लघु) attribute than the quantity of *laghu* (लघु) attribute available in the Sorghum (देवधान्य), because their glutamic acid values are much higher than that of the Sorghum (देवधान्य). In this table we can observe that vāla/pāvaṭā (वाल/पावटा), lāla cavaḻī (लाल चवळी), masura (मसुर), *mungo beans* (माष), Mung beans (मूग), and rajmaasha (राजमाष) are in possession of more values of alanine than the alanine value of Sorghum (देवधान्य). In simple way we can say that though the alanine values of these *śimbīdhānya* (शिम्बीधान्य) are higher in the given table, these can not be marked to be in possession of *laghu* (लघु) attribute in their constitution as their glutamic acid values are at more higher side than that of Sorghum (देवधान्य) , hence these *śimbīdhānya* (शिम्बीधान्य) discussed in above lines are considered to be in possession of *gururmandādi* (गुरुर्मन्दादि) attributes in comparison with the Sorghum (देवधान्य).

The average of the difference values in alanine of the selected *śimbīdhānya* (शिम्बीधान्य), displayed in the last column of this table, is 326mg/100gms. The difference between the values of actual alanine and the values of alanine generated by calculation, in more than 50% of the substances from the selected *śimbīdhānya* (शिम्बीधान्य), amazingly specifies that their values are lower than the average value of the difference (<326mg/100gms), but here we also have to consider the glutamic acid values of the concerned *śimbīdhānya* (शिम्बीधान्य). The values which are less than the average value of 326mg/100gms, are also in possession of lesser values of glutamic acid than the glutamic acid values of last five entities.

This thought discloses their susceptibility in the direction of the exhibition of the presence of *laghutīkṣṇoṣṇādi* (लघुतीक्ष्णोष्णादि) attributes in that particular entity, in comparison with the other substances of the selected *śimbīdhānya* (शिम्बीधान्य). So to some extent, here we can perceive that, if the difference value of alanine is increased, and the glutamic acid value of the concerned *śimbīdhānya* (शिम्बीधान्य) is found to be in amplified quantity, then it is possible that the concerned *śimbīdhānya* (शिम्बीधान्य),

will display some degree of excessive *gururmandādi* (गुरुर्मन्दादि) qualities.

Until now, we have clearly understood the role of glycine and alanine in the foundation of manifestation of the *laghu* (लघु) attribute. Apart from these glycine and alanine amino acids, there are some more amino acids, which also take part in the foundation of the *laghu* (लघु) attribute. We will discuss these amino acids in the second part of this series.

www.ingramcontent.com/pod-product-compliance
Lightning Source LLC
Chambersburg PA
CBHW020548220526
45463CB00006B/2226
9781692148140